D0328140

HUNTING DOWN

THE

UNIVERSE

HUNTING DOWN THE UNIVERSE

THE MISSING MASS, PRIMORDIAL BLACK HOLES, AND OTHER DARK MATTERS

Michael Hawkins

with
Celia Fitzgerald

HELIX BOOKS

PERSEUS BOOKS
Reading, Massachusetts

Property of the Library
YORK COUNTY TECHNICAL COLLEGE
112 College Dr.
Wells, Maine 04090
(207) 646-9282

Excerpts from "To a Friend Whose Work Has Come to Nothing," "There" (part iv of "Supernatural Songs"), and "A Prayer for Old Age" from *The Collected Poems of W.B. Yeats, Volume I: The Poems*, Revised and edited by Richard J. Finneran, copyright © 1934 by Macmillan Publishing Company, copyright renewed © 1962 by Bertha Georgie Yeats reprinted with the permission of Simon & Schuster. Excerpts from *Home is Where the Wind Blows* by Fred Hoyle reproduced by permission of University Science Books, Sausalito, California. Excerpts from *Tractatus Logico-philosophicus* by Ludwig Wittgenstein reproduced by permission of Routledge. "Mind" from *Things of This World* by Richard Wilbur reproduced courtesy of Harcourt Brace & Company and Faber & Faber Ltd. Excerpts from "The Nature of Action" from *Collected Poems* by Thom Gunn reproduced courtesy of Farrar, Straus & Giroux and Faber & Faber Ltd. Excerpts from *Dreams of a Final Theory* by Steven Weinberg reproduced by permission of Pantheon Books, a division of Random House, Inc. Excerpts from "Little Gidding" from *Four Quartets* by T.S. Eliot reproduced courtesy of Harcourt Brace & Company and Faber & Faber Ltd.

Many of the designations used by manufacturers and sellers to distinguish their products are claimed as trademarks. Where those designations appear in this book and Perseus Books was aware of a trademark claim, the designations have been printed in initial capital letters.

ISBN 0-7382-0037-9

Library of Congress Catalog Card Number: 98-88745

Copyright © 1997 by Michael Hawkins and Celia Fitzgerald

All rights reserved. No part of this publication may be reproduced, stored in a retrieval system, or transmitted, in any form or by any means, electronic, mechanical, photocopying, recording, or otherwise, without the prior written permission of the publisher. Printed in the United States of America.

Perseus Books is a member of the Perseus Books Group

Cover design by Dietz Design Company
Text design by Dede Cummings
Set in 11-point Sabon by Pagesetters, Inc.

1 2 3 4 5 6 7 8 9—01009998
First paperback printing, December 1998

Find Helix Books on the World Wide Web at
http://www.aw.com/gb/

TO PETER SCHNEIDER,
who took the idea seriously from the start

CONTENTS

Contents

ACKNOWLEDGMENTS

THE IDEA OF WRITING THIS BOOK was originally put to me by my agent, Margaret Hanbury, and it is a pleasure to thank her for overseeing its subsequent development.

I owe a great deal to the guidance and encouragement of Alan Samson, of Little, Brown Publishers, who had the courage and imagination to take on *Hunting Down the Universe* when it was no more than a tentative outline by a first-time author.

Many of the ideas in the book were hammered out in lively discussions with Chris Collins, Peter Brand, Alan Heavens, Andy Taylor, John Peacock, and Hugh Jones. In my customary role of devil's advocate, I all too often ended up convincing myself by my own arguments.

The scientific ideas were inspired by many friends and colleagues. In particular I thank Rachel Webster, Peter Schneider, Joachim Wambsganss, Stephen Hawking, David Schramm, Bohdan Paczynski, and Sjur Refsdal.

The book owes much of its character to discussions of the writings of other authors, but especially the works of Fred Hoyle, Steven Weinberg, and Richard Dawkins.

Caroline van den Brul of the BBC brought home to me the idea of science as above all a human activity beset with uncertainties. I thank her for inspiring me to explore the sociological and philosophical bases of the scientific process.

I am particularly grateful to Harvey MacGillivray for providing the extensive technical resources and support that I needed over many years to bring the scientific project whose results are at the heart of the book to a successful conclusion.

I received valuable help in the final stages of preparing the manuscript from Michael McCarron, Terry Fitzgerald, Catherine Hawkins, Molly Rogan, and especially Andrew Gordon, for which I thank them.

HUNTING DOWN

THE

UNIVERSE

1

CONSTRUCTING EVEREST

It is simply a logical fallacy to go from the observation that science is a social process to the conclusion that the final product, our scientific theories, is what it is because of the social and historical forces acting in this process. A party of mountain climbers may argue over the best path to the peak, and these arguments may be conditioned by the history and social structure of the expedition, but in the end either they find a good path to the peak or they do not, and when they get there they know it. (No one would give a book about mountain climbing the title Constructing Everest.) . . . It certainly feels to me that we are discovering something real in physics, something that is what it is without any regard to the social and historical conditions that allowed us to discover it.

        —STEVEN WEINBERG, *Dreams of a Final Theory*

IMAGINE YOURSELF in an airliner on a dark, clear night, not looking up at the clouds of distant stars, but downward. The sight you will see is not entirely dissimilar: a dark void irregularly interrupted by pools of light, containing millions of faint pinpricks of varying colors. If one knew nothing of the reality behind it all, what might one deduce about this mysterious universe?

With the aid of a small telescope, it would be possible to resolve the overall blur into light sources of different brightness and colors. Patterns and movements would be observed as tracers of the ebb and flow of unseen traffic. Perhaps a lighthouse would be seen as a rare luminous beacon sending out a cryptic message from the fringe of light, while ships would be made out as isolated worlds making their way across the sea of darkness.

No doubt with careful observation one could learn much about this strange place. Structures could be mapped out, motions recorded, and many correlations and patterns deduced. Subsequent discovery of superficial aspects of this universe, such as bus timetables or parts of the highway code might well appear as fundamental insights into the natural order of this world. But the idea that, underlying these islands of light and the dark voids between them was a vast globe of almost unimaginable mass, would require an inspired leap of the imagination.

In this book it will become clear that astronomers face a similar problem to that of the observer in the airliner. Until recently, nearly everything we have seen as we look at the skies has been superficial. This does not necessarily mean that it is unimportant, but the effect has been to give us a distorted idea of the nature of the Universe. However, it has gradually become apparent that underlying what we see is the vast bulk of the cosmos, which betrays itself in only the subtlest ways by its gravitational effect on familiar bodies such as stars and galaxies.

In our search for this missing mass or dark matter, we shall start by asking what science can tell us about reality and the nature of such knowledge. When we look at the philosophical basis of the scientific process, and at how astronomers seek to reach an understanding about the nature of the Universe, we shall discover a process that, far from the province of dry objectivity, is full of the stuff of life, society, and culture. We shall see science as a process of constructing order from the chaos of nature, and scientific "truth" as the provisional consensus of the community.

Great debates about the nature of the Universe have raged throughout this century, and we shall follow the way in which astronomers have fought over ideas. Out of this will emerge the idea that is the focus of the book—that the dominant bulk of the Universe exists in a form with which we can have only the most indirect and fleeting contact. The nature of this dark matter is one of the great mysteries of the twentieth century. I shall describe an observational solution that draws together many of the most exciting concepts in modern astronomy. We shall see how quasars, black holes, and big-bang cosmology unite to identify and describe the nature of the Universe's missing mass.

Toward the end of 1993, I brought to completion a project that had occupied me on and off for the previous seventeen years. The remarkable conclusion I found myself with was that about 99 percent of the material Universe is made up of primordial black holes, unimaginably dense bodies created in the turmoil of the first microseconds in the life of the Universe. This not only identified the nature and extent of the missing mass but also provided solutions to the mysteries of the mean density, shape, and ultimate fate of the Universe. But despite my belief that I had come up with a true description of reality, I cannot reasonably claim that this is the way things really are, that I had discovered something about the absolute nature of the Universe. There is no purely objective way of knowing when one is giving a true description of the external world.

I totally disagree with the sentiment expressed in the opening quote. Personal conviction is certainly not a measure of truth: the claims of religion and other forms of mysticism are probably believed with far greater certainty than any scientific idea. So what do I mean when I make such a strong claim for my discovery? In what sense is my idea scientific? What makes it a better description than, say, the equally scientific idea that the Universe is composed mainly of ordinary atomic bodies such as stars, or the unscientific belief that ultimate reality consists in the eternal realms of heaven and hell? Although we can discuss these questions in a coherent fashion, and even come up with convincing answers, it is always possible to find counterexamples to any one definition of science, and almost impossible to give a straightforward description of what it is about a scientific theory that marks it as credible or important.

When a journalist asked me whether this was the most important cosmological breakthrough of the century, I replied, "No, but if it is generally accepted it might be."[1] This answer was given a number of interpretations, all of which I think are right. They ranged from "If true, it is very important indeed" to "The truth and importance of scientific discoveries are decided by those in authority." I was trying to suggest that there are no ultimate scientific truths, or at least that there is no objective way of deciding what is really the case. We provisionally accept the validity and importance of scientific discoveries by a type of consensus. We say, or rather we should say, "This is the best idea going for the moment, but we shall have a good go at trying to

3

refute it or supplant it with a better idea," not "This is the way things really are; this is the proven truth."

Scientific theories can never be proved true beyond reasonable doubt; they can only be falsified. Even falsification is rarely clear-cut since it also relies on judgment and consensus. Eventually even the best theories are usually found to be inadequate. Like many of my colleagues I am influenced by some aspects of the work of the philosopher Karl Popper,[2] who maintained that constant doubt is the essence of scientific inquiry. There is very little room for certainty in science. Popper's ideas are not entirely prescriptive since they also describe how scientific theories actually come into being and eventually perish. It is almost an historical commonplace that great "definitive explanations" or "final theories" which everyone agrees have the inescapable feel of absolute truth are nevertheless inevitably superseded by different ideas. Brian Malpass, in his *Bluffer's Guide to Science*, advises that

> if ever a truly great man of science . . . pronounces with great conviction that something is impossible, it is a safe bet he will be proved wrong shortly. And should another equally great man of science rush to endorse his colleague's assertion, you are advised to mortgage the family dwelling and hasten to a betting shop to invest the proceeds at two to one and await the imminent announcement to the contrary.[3]

But why should this be so?

There might be some disagreement about the best way to climb Everest, but surely, as Steven Weinberg asserts in the opening quote to this chapter, we know when we have achieved our objective or at least made some progress toward it. Weinberg assumes that scientific activity is analogous to mountain climbing, where the only purpose is to reach a known destination, but even mountain climbers might challenge this description of their purpose. There are easier ways of reaching mountain peaks. Like a long-distance truck driver I encountered in Chile, mountain climbers might claim that their road is their destination. (The driver had the words *El camino es mi destinado* painted on his truck.) If you know the destination, then there is no point in looking for it. If you already know what the truth looks like, then why try to find it? Scientists are more like mountain climbers who have never climbed real mountains, and are not only uncertain of their existence but have constant misgivings about the validity of their collective activities. Their mountains are imaginary. They are mental constructs where the only checks on the imagination are consistency with agreed procedures such as compliance with mathematical coherence and what counts as observational evidence. Even fundamental requirements such as these are subject to debate and evolution as mountain climbing techniques change. Thus the evolution of the strategy of mountain climbing gives rise to the gradually shifting idea of mountains. So scientific activity can be thought of as analogous to constructing Everest, rather than discovering or surmounting it, but I think the idea of evolving Everest is more appropriate. Science is not an edifice; it is a process. Scientists are human, and along with all their activities, ideas, and constructs, they are biological entities whose enormously complex existence can only be explained in terms of Darwinian evolution.[4] Just as in biological evolution, deviations from the generally accepted ideas or the "standard models" of science cannot be too radical without risking almost certain death from a hostile environment. Conformity with accepted models may be one of the criteria for the survival of a scientific idea, but this is not the same as saying it is true.

All that we can reasonably be sure of is that human beings are more intelligent than the Universe. Nature is mindless, and is in itself essentially incoherent and chaotic. It merely provides the raw material for our ideas of the way things are. Just like any life form, reality represents organization: it is the product of our combined intellects in bestowing order on nature. Therefore it reflects our psychological makeup and our social and political structures, which are all manifestations of the collective evolution of humanity. Hence reality is a constantly evolving entity. But it is not evolving toward anything, any more than an organism can be said to be evolving toward an ideal life form. Every organism has its own coherence and validity. The very fact of its existence is sufficient reason for its being. Just as there is no purpose to evolution, so there is no "correct" idea of reality. The road is the destination. Scientific endeavor is no more a journey toward absolute truth than biology in all its diversity is a set of constantly improving stages toward some state of ultimate perfection.

Hence scientists are never right, they just have a more or less strong case, which they argue with varying degrees of skill, determination, and conviction. Science is a human activity. It is a set of games that is rule bound, competitive, and has the agreed objective of describing the nature of reality. These rules and this objective are all that might be regarded as constant or certain, although even they can be modified over time by evolutionary processes. Scientists are no better placed to determine ultimate reality than philosophers or mystics. They do not have some special perspective for accessing the truth. The search for reality is an interminable point-scoring activity. The most we can hope for is to get better at playing the games of science.

The ostensible purpose of these games is to understand the material Universe. This is the complex objective toward which we see ourselves as progressing by a mixture of leaps of the imagination, argument, and rigorous testing. The scientific process is essentially creative. We often arrive at good theories by inspiration, and coming up with good ideas is the enjoyable part of the scientific process. The real challenge of making conceptual breakthroughs starts when the idea is submitted to tests for acceptability by oneself and the scientific community. Many beautiful scientific insights have been relegated to obscurity because they failed early attempts to falsify them or were shown to be inconsistent with the existing framework of accepted theories. What we think of as scientific facts or convincing theories are just those insights that nobody has as yet either convincingly falsified by observational evidence or supplanted with a more beautiful, robust, and imaginative idea. "Scientific truths" is simply another way of saying "the fittest, most beautiful, and most elegant survivors of scientific debate and testing."

As an orthodox scientist, I have more faith in the scientific approach to understanding the Universe than in religious or philosophical methods. This is because, at least in principle, scientific theories are always open to argument, scientific proofs can be falsified by evidence, and evidence is always open to interpretation. Science is exposed to a more exacting and dynamic hostile environment than absolutist systems such as Christianity. I am suspicious of any ontological system that claims to deliver unchallengeable truths. The extent to which scientists claim to have delivered such certainty is the extent to which they have perverted the real purpose of science, which is above all a rigorous but open-minded and dynamic system of inquiry.

To proceed with the task of trying to examine and discuss the way things are in the external world, we must make at least three assumptions. We take it as undeniable that there is an objective, physical reality; that we can in principle discover the nature of this reality; and that the only way to do so is by the scientific method. These assumptions are regarded as unassailable, and are not open to debate. All systems need some unshakable premises. If these assumptions were called into question, it would become a different game. We simply have to believe that we are examining and trying to describe the real world as it exists in itself. Thus most of us hold as a matter of faith the idea that the Universe is the way it is regardless of our perception of it. Scientists can be wrong about the way things are, but for the purpose of playing the game of science they cannot doubt that there is objective truth about which one can be wrong. (However, as we shall see in Chapter 7, quantum mechanics, the study of matter at the atomic level, provides astonishing and important exceptions to all this in that it does call such assumptions into question.) The problem is how to appreciate the extent to which we are influenced by nature and the extent to which we are simply playing an internally consistent game in which the so-called external evidence is made to fit our elaborately constructed theories. The answers to this problem are controversial and far from simple, and I hope to examine them throughout the book.

Scientific truths are determined by a sort of consensus. All the same, we behave as if such truths are decided by external evidence. Many scientists really do believe that they have a hotline to the truth. What is more, they have persuaded most of our population that scientists have this godlike wisdom and that what is scientific must be right. Scientific evidence is often considered to be beyond challenge by, for instance, courtroom juries. Not only are scientists seen as dealing in absolute truths, but many people think of the scientific community as being entirely objective and almost infallible. This is not only a fallacy but, I think, rather boring. The scientific process is far more subjective and uncertain and so much more interesting.

Rather than discussing the history of so-called scientific facts, which we can get from any good encyclopedia, I hope instead to examine the way we go about the business of science or, at least, the way I think we should see the scientific process. I also hope to describe specific examples of how astronomers go about the business of making discoveries, how these are received by the astronomical community, and the tortuous process of trying to obtain acceptance for our ideas. Although I am unlikely to be entirely objective about the status and progress of my own ideas, I am on the other hand familiar with the process of developing them and trying to persuade my colleagues and the world at large of their virtues. The theory which I have put forward, that the material Universe is almost entirely made up of small black holes, is, I think, sufficiently novel and provocative to provide an interesting case study of the conception, birth, and struggle for survival of a new scientific idea. Consequently, this is the main focus of the book. I hope to persuade you that my solution to the dark matter problem is a really good and exciting idea, as well as one with strong supporting evidence. However, the most I can demonstrate is that my idea works very well within, and is a natural extension of, our current model of the Universe, and that it not only arises from but adds enormously to the robustness of our current mainstream cosmological theories.

Science is a fascinating human drama. It is a set of games that anyone can join in at some level provided they are given access to the rules and can see the process taking place. Consider how unsatisfactory the institution of soccer would be to noncombatants if the only information the soccer authorities thought we could handle was the score. What if we were not entirely put off and bored by this state of affairs and asked for more information and were then given a whole lot of confused metaphysical guff involving inexplicable forces and existentially ambiguous possibilities? Obviously this would make us suspicious. What would we deduce from this? Would we think, "This is clearly beyond us and should be left to the experts," or "These people must be crazy"? Perhaps we would conclude that everyone involved in soccer wanted to keep it all veiled in mystery and was feeding us meaningless information to convince us that we are too stupid to evaluate the way the game is played. Perhaps they believe that, like the Delphic Oracle or the British royal family, their power and possibly their very existence depend on perpetuating the myth that their tasks can be performed and correctly evaluated only by innately superior beings, and that a shield of enigma is necessary to protect them from the dangerous scrutiny of ordinary mortals. Whatever form our suspicions might take, we would certainly feel either insulted or humiliated by this type of response.

Although I do not believe that the scientific community deliberately sets out to withhold information or to mystify the general public, there is nevertheless a large element of exclusivity and secretiveness about the scientific process. This elitism is especially noticeable in the fields of particle physics and astrophysics. For instance, many astronomers, especially the theoreticians, feel that their subject is all far too complex for the general public to grasp. They feel comfortable only when publicizing cut, dried, packaged, and approved ideas. When questioned about their reluctance to allow the general public access to anything but the most trivial and technical aspects of the scientific process, they will say that they are afraid of giving the public the wrong impression, that people will be confused by the uncertainty of it all, that only trained astrophysicists can understand and live with the whole complicated business. To me this sounds not only arrogant but absurd. All of us are constantly bombarded by the media with the far more complex, confusing, and slippery business of domestic and world politics. Most of us can handle that. We are all involved in the process of politics, even though we are not politicians. Perhaps some scientists are afraid that people may notice their feet of clay and lose respect for them. Maybe they genuinely believe in their own superiority.

The opening quote of this chapter suggests that Weinberg thinks of scientists as possessed of this sort of omniscience. Apparently they are blessed with a superior instinct for recognizing truth. Also, "It certainly feels to [him] that [scientists] are discovering something real in physics."[5] This quote is taken from a chapter where Weinberg expresses his contempt for philosophy, sociology, and anthropology, which he seems to regard as the enemies of modern science. It is part of his response to ideas such as those contained in Andrew Pickering's book *Constructing Quarks,* and work by Bruno Latour and Steve Woolgar, who maintain that "the negotiations as to what counts as [scientific] proof or what constitutes a good assay are no more or less disorderly than any argument between lawyers and politicians."[6]

I see the sentiments expressed by Weinberg as representative of the attitude of the theoretical physics establishment. To my mind, the opening quote exactly expresses the root cause of science's alienation from the public it is meant to serve and whose money makes it possible for physicists to spend a billion dollars looking for the top quark. The first question we should ask is why Weinberg considers sociological, anthropological, or psychological explanations for the scientific process so repugnant. Does he believe that scientists somehow transcend their historical context and their humanity, that the nature of their quest is so otherworldly that they have to obtain a sort of transcendent inhumanity in order to be in the state of grace that qualifies them to know when "one view or another bears an unmistakable mark of objective success"?[7]

On the contrary, I believe that scientists are all too human and that many of us just cannot come to terms with the uncertainty and flimsy bases of many of our scientific beliefs and take refuge in authoritarianism and the ivory tower of intellectual elitism. In consequence, many attempts to explain scientific ideas will often seem incomprehensible. This is not because the ideas or explanations are too difficult to understand, but because they are in themselves often confused, unbelievable, and sometimes do not make sense. Also, we often overlook the extent to which the nature of the evidence depends on theory, and on the way experiments are conducted and interpreted. We are also largely unaware of the ontological limitations of formal logic or mathematical physics which help us to shape and evaluate our theories.

Thus many scientific theories appear to fly in the face of common sense, and a common reaction among scientists is to consider them beyond the reach of anyone but a specialist. This attitude makes for a mystique surrounding scientific ideas which is entirely unjustified. There are few if any scientific concepts that cannot be explained and made believable or revealed as nonsense to an interested layman.

Hopefully, the truth of this will be shown in the chapters that discuss the problem of missing mass or dark matter and my solution to this enigma. It should demonstrate the essential simplicity and accessibility of what might seem like some of the most remote and intractable ideas at the cutting edge of cosmology.

But before embarking on this, so that my work can be seen in context, I want to explore the main issues underlying our picture of the Universe. Discussing the great cosmological debates of our century such as the struggle between proponents of the steady-state theory of the Universe[8] and big-bang cosmology,[9] as well as the fallout from and developments of these issues, should supply a pretty comprehensive idea of what modern cosmology is all about.

I also wish to examine the conflict between rationalist and empirical approaches to science. I think that this philosophical disagreement, which hinges on a fundamental difference of opinion as to whether the route to truth is through the intellect or through experience, is crucial to an understanding of the scientific process. Among other things, it enables us to appreciate what is really behind put-down remarks like "reality cannot be understood in terms of ordinary common sense" or assertions such as "tense is not real." This struggle for supremacy between theory and evidence in shaping our idea of reality is explored in the chapters entitled "Reality in the Dark" and "In the Land of the Blind." Meanwhile, we will go straight to the dramatic and passionate story of cosmological and philosophical conflict between two titans of modern astronomy.

2

DESCENT IN THE MIST

Big-bang cosmology is a form of religious fundamentalism, as is the furor over black holes, and this is why these peculiar states of mind have flourished so strongly over the past quarter century. It is in the nature of fundamentalism that it should contain a powerful streak of irrationality and that it should not relate, in a verifiable, practical way, to the everyday world. It is also necessary for a fundamentalist belief that it should permit the emergence of gurus, whose pronouncements can be widely reported and pondered on endlessly—endlessly for the reason that they contain nothing of substance, so that it would take an eternity of time to distill even one drop of sense from them. Big-bang cosmology refers to an epoch that cannot be reached by any form of astronomy, and, in more than three decades, it has not produced a single successful prediction.
          —FRED HOYLE, *Home Is Where the Wind Blows*

> Should such a man, too fond to rule alone,
> Bear, like the Turk, no brother near the throne,
> View him with scornful, yet with jealous eyes,
> And hate for Arts that caused himself to rise;
> Damn with faint praise, assent with civil leer,
> And without sneering teach the rest to sneer.
>   —ALEXANDER POPE, *"Epistle to Dr. Arbuthnot"*

IN CONTRAST to the archetypal big-banger Steven Weinberg, who compares science to ascending Everest, Fred Hoyle, the more down to earth steady-statesman, draws a powerful metaphor for scientific endeavor from his experience of climbing the few hundred highest hills of Scotland, popularly known as the Munros. Although both

13

Everest and the Munros are fraught with danger from sudden changes in the weather, almost anybody can climb the Munros, whereas only a select few heavily sponsored mountaineering experts can hope to ascend Everest. Everest is analogous to the idea of cosmology as a great and almost inaccessible edifice that can be tackled only by single-minded teams with extensive and expensive support structures; the Munros may be likened to cosmology as a series of empirically testable, individual achievements by small groups or solitary explorers with relatively modest resources and a good attitude. Taken together, these individualistic achievements eventually delineate the larger landscape, and in a reliable way in that the hill walkers' experiences can easily be verified or falsified. Thus the larger picture can be drawn from known facts rather than from a set of mutually sustaining hypotheses that are needed to keep the more ambitious high-stake mountaineers moving together along a predetermined path that, according to Hoyle, most probably leads nowhere. In his experience of hill walking in parties, when a mist suddenly obscures the way down there is invariably one member of the group who confidently sets out for home, with nearly everybody else blindly following. These faithful followers are always led completely astray. If there are one or two who are more skeptical of the self-appointed leader's conviction that he has a "feel" for the right way, and they decide to follow their own course based on tangible, piecemeal evidence, inching their way patiently through the mist, they will usually arrive home long before the main group.[1]

In the examination of the philosophical conflict between rationalism and empiricism in Chapter 4, I have characterized Hoyle as a "Platonic realist," which is another term for "rationalist." As one might expect from a theoretician, and especially a cosmologist, he is, in important ways, just that. His belief that there has to be something like a divine purpose to everything is especially evident in his theory of the origin of life, which opposes the atheistic implications of Darwinism.[2] Also, like Steven Weinberg, he believes that good scientists are discovering something about the real nature of the Universe, and that it is just a matter of finding the right path (applying the correct methodology) to uncover incontrovertible facts. Nevertheless, his early training was in quantum physics, and his gurus were staunch empiricists who placed heavy reliance on experimental and observational evidence. Therefore his ideas of correct scientific methodology are, up to a point, strictly empirical, as his lost-in-the-mist metaphor indicates. Unlike Weinberg, Hoyle does not believe that scientists have an instinct for the "correct" way. The feeling of "knowing when one view or another bears the unmistakable mark of objective success" is entirely unreliable. He is perhaps the first great empirical cosmologist with a repugnance for systems of unsubstantiated hypotheses, which he regards as little better than flights of fancy. Like religious dogma, such ideas are unfalsifiable by observational evidence and therefore, like all unscientific systems, can be believed, often passionately, only as a matter of blind faith. This inevitably leads to a type of fundamentalism entailing intolerance and eventual inertia, where new ideas are nipped in the bud and students and other initiates are required to be uncritically subservient to the teachings of their aging gurus.[3]

Indeed, it takes almost suicidal courage to leave the herd and challenge the authority of the astrophysical establishment. Typically, papers expressing genuinely new ideas are refused publication by referees of reputable scientific journals on the grounds that they undermine the generally accepted principles of physics. Those who persist in writing such papers are usually sidelined from the astronomical community by their peers. Admittedly, most of these new ideas are highly implausible, and usually turn out to be wrong in the end. The problem in my view is the manner of their dismissal, which is usually by a vague appeal to general principles. Often such challenges to the mainstream picture are based on observations that appear to contradict generally accepted ideas. The challengers seem to feel, in the spirit of Karl Popper, that it is sufficient to find a contradictory observation in order to refute a theory. In reality this is hardly ever adequate, for two main reasons. First, observations themselves can only be interpreted as part of a network of other theories, and many pitfalls can lie between photons being gathered by a telescope and the unambiguous conflict with the predictions of a theory. Second, and, usually, much more importantly, theories are rarely, if ever, rejected until there is a more satisfactory replacement. New ideas that rely on discrepant observations are never taken seriously unless they are underpinned by a theory that explains them in a natural way, and makes further testable predictions. If new ideas were criticized on this basis by the establishment, rather than by disdainful dismissal, we would, I feel, find ourselves in a much livelier and more creative astronomical scene.

In my early years as an astronomer, the ultimate heresy was to give comfort and support to the steady-state cosmology of Fred Hoyle, which opposed the big-bang dogma of the establishment. Big-bang cosmology is the idea that the entire Universe is a system that came into being in a single explosive instant that caused it to expand and evolve so that it was different at different stages of its history. Because of the finite speed of light, when we look out into space we are also looking back in time, so the Universe looks different at different distances from the Earth. The major weakness of the original big-bang cosmology was its inability to account for the fact that at 10^{-43} of a second after the origin of the entire Universe, the balance between the expansionary force of creation and the contractive force of gravity had to be infinitesimally fine. In other words, there had to be just enough matter to hold the Universe together in dynamic equilibrium. It must have started out with exactly this critical density in order for it to be, in cosmological terms, so close to the critical density today. This critical balance between expansion and contraction must have been to an accuracy of 1 part in 10^{60}—that is, 1 in 1,000,000,000,000,000,000,000,000,000,000,000,000,000,000,000, 000,000,000,000,000. Without such a fine balance, the Universe would long ago have collapsed into a black hole or expanded to such an extent that it would now be virtually empty.

According to the steady-state theory, the Universe is infinite in time and space and therefore, on a sufficiently large scale, it is the same everywhere and at all times. The observed expansion is explained by the continuous creation of matter everywhere. This also keeps the mean density constant, so there is no need for the finely balanced initial conditions that can only be accounted for by a metaphysical appeal to divine intervention. After 1985, when particle physics had advanced sufficiently to provide the physical mechanisms, Hoyle produced a revised steady-state model in which intermittent bursts of creation occurred within "C-fields."[4] But originally the idea was of a constant flow of creation which, though mathematically coherent in accordance with Einstein's theory of general relativity, was ultimately physically unsustainable.

Around 1980, shortly after the arrival of our new director at Edinburgh's Royal Observatory, we had a talk from Jan Einasto, an influential cosmologist who had been working on problems associated with the large-scale structure of the Universe. I asked him whether his model of a honeycomb structure was consistent with the steady state. To my astonishment, the director turned around in his seat and glared at me for a full fifteen seconds. Such talk was not to be tolerated. The big bang had been established as the favored theory and should form the basis of any cosmological discussion. After this I found that any research that was not directly related to the director's view of the establishment agenda was discouraged. Nevertheless, he confided his disgust at the discovery that the *apparatchiks* in the totalitarian Soviet Union had issued an edict concerning the favored status of the steady-state theory. In the Soviet Union the official cosmology was Hoyle's theory, while in the West, as sanctioned by the Vatican and other authorities, the official theory was the big bang. At that time, one of the more wayward thinkers at the Royal Observatory, Victor Clube, was sidelined and transferred to a government sponsored post at Oxford University, on the grounds that his ideas conflicted with some fundamental tenets of modern physics. I survived as an astronomer, but only just, because as an observer rather than a theoretician, my usefulness lay in producing all sky catalogues available for anyone to use, rather than pushing ideologically loaded ideas. Also, although I enjoyed arguing the pros and cons of the steady-state theory, it was evident to me that it could not survive in its simplest form. What annoyed me then, and still does, was the way in which the steady-state theory was dismissed. The worst I could be accused of was irreverence and lack of gravitas, but not actual heresy. I am far from being antiestablishment. It is just that I do not see any scientific ideas as irrefutable, and regard them as untrue only if they are blatantly at odds with unambiguous observational evidence. However, there is very little of that in present-day cosmology. Almost all evidence is open to a number of interpretations, and much of it is suspect anyway.

If all this is not an example of religious or, at least, ideological fundamentalism, then what is it? It is difficult to imagine clearer support for Hoyle's eloquent condemnation of this state of affairs quoted at the beginning of this chapter. To be fair, the ideological fundamentalism was a little less overt than has been implied. Our director was engaged in a power struggle for the privilege of having

Edinburgh as the European headquarters of the future Hubble Space Telescope. He could not afford to have politically naive loose cannons in his establishment. But he was undoubtedly a strict big-banger who disapproved of dissenters, and so he was not being entirely cynical. His political astuteness merely made him more overtly repressive, since he was only too aware that his views were shared by the wider astronomical establishment who would be judging Edinburgh's suitability as the European administrative home for the prestigious space telescope. He also wanted Edinburgh to be seen as a good service establishment, with only a few hand-picked and closely controlled research workers, not a bunch of self-directed scientists pursuing their own agendas. This was at least partly because he wished to comply with the mission of government scientific establishments in the United Kingdom, which was revised in the early 1970s with the prime objective of serving the expensive practical needs of the universities. Inevitably, as with all politically created entities, the scientific civil service turned into a Frankenstein's monster with a will of its own. Instead of merely administering and developing the technology to be used by university researchers, it started to mimic and consequently rival university departments but with the immense advantage of controlling the technology and hence the direction of research. At the time there was pressure on the United Kingdom's Science Research Council (now called Particle Physics and Astronomy Research Council or PPARC) to curb this tendency and give the universities a fair go. All this explains why our director discouraged research work by government scientists.

Fred Hoyle alleges, with some justice, that there is a terror of the new in British astronomy. When I was a postgraduate student at Cambridge University after completing my mathematics degree at Oxford, some three years after Hoyle had set up the government-sponsored Institute of Theoretical Astronomy at Cambridge, I was deeply impressed by the moribund atmosphere of the Observatories (the university's astronomy department). Martin Ryle's radio astronomy branch of the physics department at the Cavendish Laboratory was the flagship of British astronomical achievement at that time, and was practically a no go area for members of the Observatories.

Hoyle dedicated much of his professional life to improving astronomy at Cambridge, for which he was rewarded with public humiliation engineered by Martin Ryle, and eventually with a statue in the Institute's garden. At first, Hoyle and his group of theoreticians at his government-sponsored Institute were fairly remote from the rest of the university since it was not in their mandate to teach. They produced excellent, far-reaching theoretical work under the envious eyes of the astronomy department and Ryle's Cavendish radio astronomy group. Hoyle claims that the Cavendish, which was heavily sponsored by the Mullard Company (an electronics/radio company), was money guzzling, secretive, jealously guarded, and third rate.[5]

Fortunately for me, the astronomy department reacted to the Institute's success by poaching, as its new head, one of the best British astronomers. Donald Lyndon-Bell was a scientific civil servant at the Royal Greenwich Observatory, the government institution then based at Herstmonceux Castle in Sussex. He had an awesome reputation for mathematical insight and was universally respected. As soon as he was appointed, he set about clearing the deadwood from the department with much determination but only moderate success, though at least one member of Cambridge's astronomy department was driven into the church and another to drink. As a highly principled, imaginative, and energetic astronomer, he was an icy blast of fresh air. To me he was the ideal role model. He gave me invaluable insight into the way to approach scientific problems, and pointed me in the right direction, which was as far from the claustrophobic, blood-drenched arena of Cambridge as I could possibly go. Although I was still officially a Cambridge student with Lyndon-Bell as my supervisor, and though I eventually obtained my Ph.D. from Cambridge, I spent a year at Cape Town under the benevolent eye of Richard Woolley, a former director of the Royal Greenwich Observatory and retired Astronomer Royal. I completed my doctoral thesis while at Edinburgh University, whose astronomy department is closely associated with the Royal Observatory. It was to their mutual benefit, not least because it avoided damaging rivalry, that the two establishments operated under a single head who was both observatory director and university professor and, until quite recently, Astronomer Royal for Scotland. Consequently, although it does not have the deceptive allure of high profile names like "Cambridge," and although, as I have indicated, it was not entirely immune from the ubiquitous Cambridge disease which to some extent infected most of the world's astronomical community,

Edinburgh has nevertheless become one of the best centers for astronomy in the world.

Fred Hoyle should be kicking himself that he turned down the Scottish Office's invitation to locate his Institute at Edinburgh or Glasgow. He did this on the grounds that neither city was Aberdeen, which, being close to his beloved Munros, was his second choice after he was initially refused Cambridge. If his center of excellence had been located in Scotland, he would probably not have had to fend off the Cambridge astronomy department's pervasive miasma and stultifying traditionalism, and the subsequent painful culture of intolerance among the survivors of Lyndon-Bell's remedial measures. Nor would he have incurred the jealous wrath of his close rival, Martin Ryle, who, in the words of Alexander Pope, "too fond to rule alone, [could] Bear, like the Turk, no brother near the throne." The whole course of cosmology could have taken a different turn, or at least would have had a different complexion. It may be hard to believe, but cosmology really is a seriously emotional issue, and can arouse irrational passions in otherwise intelligent and sensible men and women. I very much doubt that this would have been the case if it were not for the inconclusive bloody battle between Hoyle and Ryle which was fought out in the Cambridge arena. Although it had international repercussions with participation from astronomers all over the world, and although they employed scientific methodology and observational evidence as weapons, the struggle was above all a parochial battle for territorial supremacy between a couple of ex-Admiralty, Second World War radar boffins with a long-standing professional rivalry and immense personal influence. They each controlled huge budgets, provided by the government in Fred Hoyle's case and by the private sector in Martin Ryle's.

Traditionally, Britain has been regarded as the leader in theoretical astronomy, while the United States had sufficient resources to lead the way in observational work. Consequently, most of the important discoveries have been made by astronomers working in America although our present-day understanding of the Universe is largely due to theoreticians working at Cambridge.

The influence of Fred Hoyle on modern cosmology cannot be overestimated. He even invented the term "big bang," which he offered originally as a disparaging remark. Although their differences have been blown up all out of proportion to the scientific issues at stake, the two theories represent only slightly different attempts to address the same problems, many of which have still not been resolved. Their main difference was, and still is, ideological and, in some ways, even philosophical. The sticking point has always been Fred Hoyle's refusal to countenance the biblical idea of the Universe coming into being from nothing. Although he is a Christian, his idea of God is that of a far more human entity than the awesome, all powerful figure of the Old Testament. To create absolutely everything, including the laws of physics, is just too much to expect, even from God. Although Hoyle's line of reasoning may sometimes come very close to argument from personal incredulity, he nevertheless makes the very good point that if one has two competing ideas, then surely one should choose the more believable, the one that makes the most sense and comes closest to being testable, and certainly not the one that makes a mockery of the whole business of empirical science by proposing an inaccessible state of affairs, such as the big bang or black holes, where the laws of physics no longer apply.

The mechanism for coming into being in the big-bang creation picture is as profoundly imponderable as the mechanism for God's creation of the Universe. If there are no physical laws in operation, then there is nothing for scientists to investigate or describe. Thus Hoyle sees big-bang cosmology as essentially unscientific in that, contrary to the prescriptions of Karl Popper, it is unfalsifiable since it is little more than a system of mutually sustaining hypotheses. The strength of the steady-state theory is that it is not only inherently testable, but it makes specific falsifiable predictions. It has no need for speculative physical conditions such as hyperspace or black holes. However, unlike the indestructible, multiheaded hydra of big-bang cosmology, the steady state is, by its nature, vulnerable to empirical evidence. In contrast, nothing, apart from a change of heart, can unseat a metaphysical theory such as the big bang. No matter how many times one of its heads is chopped off, a similar one will grow in its place. As soon as one hypothesis is discredited or starts to look insecure, another equally speculative replacement is pulled out of the hat. When everything starts to get too complicated and, like the protagonists in a Greek drama, the big-bangers find themselves inescapably entangled in a conceptual impasse, they can always appeal to the deus ex machina of the imponderable initial conditions of the creation. If they cannot prevail on the exposed battlefield of physics, then they can beat a retreat into the impregnable fortress of metaphysics. This structure is buttressed not only by the submissive astrophysical majority, but also by tame journalists and broadcasters. Such stakeholders in the pseudoscientific creation myth help the astronomical establishment to disseminate what amounts to the sensationalized propaganda of big-bang cosmology, in order to impress the public, whose money makes it possible to spend countless billions on projects like the Hubble Space Telescope.

All this, as I understand it, is Fred Hoyle's very strong view on the intellectual impoverishment of big-bang cosmology and its regressive social consequences. He clearly believes that the choice of imponderable and therefore scientifically inaccessible initial conditions of creation is not just a matter of a local disagreement among astronomers. It is absolutely fundamental to the whole philosophy of science, and has serious impacts on all of society, especially on the intellectual freedom of scientists. A system that depends on faith leads to censorship and denunciation rather than reasoned argument. This must be socially undesirable. A belief set that persecutes and ridicules those who refuse to blindly submit has to be positively rotten.

Most of the individuals involved in this intractable dispute are now dead or inactive. Nevertheless, departures from accepted standard models are still sometimes seen as threats to the established view rather than as potentially interesting new developments. I recently experienced some angry denunciation and sneering disapproval for having the temerity to come up with and publicize my new cosmological idea. Admittedly, I was in no way trying to shake the foundations of big-bang cosmology, but my impression is that such reactionary relics from the 1970s and 1980s are fairly rare. The vast majority of the international community of astronomers, though skeptical, as they should be, inquired into my work in the spirit of true scientific curiosity. Many of them, without knowing or caring whether it was right, expressed immediate interest in what they saw as a good debatable idea.

The influence of Cambridge on British and international astronomy has steadily waned since Hoyle and his group disbanded and the Institute merged with the university. Ironically, Cambridge's influence depended largely on the formidable reputation of Hoyle and his group of theoreticians. Cambridge University used this borrowed glory to seize and kill the same golden goose that had lent them their reputation and enriched British astronomy as a whole. In this way they impoverished not only their own establishment but the whole of theoretical astronomy. Unchanged, of course, was Cambridge's medieval culture of elitist traditionalism and fundamentalism. Because there is always a certain amount of inertia that enables things to roll on well after they have been set in motion, Cambridge continued to exert enough influence to enable the successful export of this culture to the wider astronomical community. The orthodoxy of big-bang cosmology is essentially a creature of Cambridge culture. Consequently, as Cambridge astronomy's influence waned, so did the unquestioning acceptance of the big bang and the related business of truth by decree. Astronomers are now much freer to pursue their own ideas, with the result that astronomy and therefore our concept of the Universe has become more dynamic and debatable. This is the way it should have been all along instead of the species of religious fundamentalism that it has been for decades and, to some extent, still is.

Having thus set the stage for the enactment of the contest between big-bangers and steady-statesmen will perhaps make the fury and passion of the whole drama more understandable than an unadorned discussion of the scientific issues could possibly do. To anyone without some appreciation of the truly medieval nature of Cambridge, Hoyle's virtual excommunication from the astrophysical community for possibly being wrong over one aspect of his steady-state theory would be too incredible. We can believe that the church did something similar to Galileo, but not that a twentieth-century establishment did it to Britain's greatest postwar astronomer. We can even sympathize with Galileo's persecutors. His heliocentric model of the Universe was, after all, a challenge to widely held convictions about the central role of man in the divine order. He was shaking the very foundations of painfully acquired Western civilization and threatening to cast our ancestors into an intellectual wilderness without purpose or order. There is no excuse for what was done to Hoyle.

However, the main difference between the big-bang and steady-state cosmologies is, when it comes down to it, philosophical. It all hinges on the acceptability of a single moment of creation. Hoyle sees the idea of a beginning to everything as fundamentally incoherent. The scientific differences, which were never all that significant, have been steadily eroded as both theories have been modified. They are almost unrecognizable as the original rival models of the Universe, and have all but converged. Fred Hoyle now sees the Universe as an infinite collection of little bangs, while big-bangers now speculate that "our" big bang may be part of a larger manifold in which big bangs are created under suitably favorable conditions. In a way this is a cop-out for Hoyle since the multiple-little-bang model, unlike his earlier versions of the steady-state theory, is empirically inaccessible. Such reliance on a set of ideologically driven hypotheses to sustain the empirically insupportable has always been one of his main criticisms of the standard big-bang theory. However, there is still nothing in his theory to encourage the fashion for pseudoscientific speculation about creation. For Hoyle, the Universe is in principle empirically knowable; we are simply not in a position to know all of it. Since the lack of a beginning to time is what distinguishes Hoyle's steady-state theory from big-bang cosmology, he has always avoided what to him is the philosophically unacceptable idea of everything being created out of nothing, and the most important aspect of steady-state cosmology remains intact. Although it might seem that Hoyle has capitulated in all but name, he has not given an inch on the pivotal reason for the conception of the steady-state theory. Consequently he sees it as being as subversive to big-bang cosmology as it has always been. Hoyle has no truck with the idea that modern developments in the big-bang theory are converging with the steady-state theory.

The bottom line is that Hoyle sees the big-bang/steady-state struggle as a continuing battle between physics and metaphysics in cosmology. Although it is important to appreciate his point of view, and to understand the historical context that shaped his extreme attitude to the astronomical establishment and its beliefs, I think it is far too simplistic to categorize big-bang cosmology as metaphysics just because it finds itself up against imponderables and is occasionally a bit too imaginative when proposing solutions. There is an element of truth in Hoyle's characterization of big-bang cosmology. For instance, some descriptions of black holes verge on the ludicrous and are certainly beyond empirical science. Nevertheless, although we cannot precisely describe them without engaging in a type of metaphysical speculation, there is very little doubt that something like black holes must exist in order to make sense of our current understanding of the physical laws governing the Universe. Also, contrary to Hoyle's claims, their existence is as empirical as any other astronomical phenomenon. We know how to go about observing them. Such observations may be indirect and theory dependent, but this is true of nearly all scientific evidence, especially in astronomy. He is also somewhat inconsistent. He praises a theory for being imaginative and leading to genuinely new ideas, and in the same breath condemns another for being too fanciful and outrageously speculative, but nevertheless trivial and unexciting. This looks very like a straightforward value judgment based on personal preference.

The received wisdom about Hoyle's steady-state theory is that it might be elegant and beautiful, but it cannot withstand the harsh reality of empirical evidence. Both sides in the struggle appeal to Popper's scientific prescriptions in order to discredit each other. The big-bangers claim that Hoyle's theory has been falsified by the evidence, while Hoyle claims that the big bang is unfalsifiable and therefore unscientific. I do not believe that either side is right. I also think that Popper's view of science is an oversimplification. Apparently falsifying evidence is by no means necessarily sufficient to discredit a theory. The evidence may itself be suspect and its interpretation dependent on the preconceptions of a rival theory. Also, as Hoyle points out in a statement inconsistent with his view of the big bang as metaphysics, there has been a great deal of evidence that could easily have been used to unseat big-bang cosmology if there had been enough collective will to do so. Because the establishment and, therefore, the majority of astronomers subscribe to big-bang cosmology, it has been repeatedly forgiven its mistakes, whereas the steady-state theory was never allowed even the slightest benefit of the doubt.

As you may appreciate by now, it is almost impossible to disentangle the purely scientific arguments from the political, social, cultural, philosophical, and personal issues that go into shaping our idea of the Universe. In some ways the so-called scientific facts are the least interesting part of any cosmological theory. The process of abstracting them from the dynamic environment of their conception and development extinguishes their vitality. Facts are like stuffed animals in a glass case, only remotely suggesting the wild uncertain environment in which they had their being and which was an inextricable part of their nature. Nevertheless, the scientific arguments are obviously important in themselves, and now that the general scene has been set, we can explore the substance of the big-bang/steady-state struggle in greater detail, and look more closely at some of the events only alluded to so far.

3

SURVIVAL OF THE WEAKEST

A major reason for the popularity of big-bang theory is undoubtedly that it is simple enough to place no burden on the mind. Undoubtedly, too, there are many who are attracted by its retreat into metaphysics. For myself, I find the retreat into nonexplanation unsatisfactory, contrasting so markedly with the exquisite subtlety of all science outside cosmology. Can the Universe really be so crude while all the rest is so refined?

— FRED HOYLE, *Home Is Where the Wind Blows*

If a footballer makes ten mistakes but scores two goals, he is king of the night.

— LUCIANO PAVAROTTI

WHAT ARE THE SCIENTIFIC insights that underpinned big-bang cosmology and the steady-state theory? It all boils down to the need to explain why the force of gravity has not pulled everything in the Universe together into one infinitely dense entity, a singularity. This, coupled with Edwin Hubble's gradual realization in the late 1920s that distant galaxies are receding from us, led to the idea that the Universe is expanding, and that the reason for the expansion is the immense energy of creation acting as a counterforce to gravity. The big bang and the steady state are, essentially, rival creation theories, which in different ways explain why the Universe is in the dynamic equilibrium between the forces of expansion and contraction which has to be the case in order for it to have existed for more than an instant, let alone for billions of years and possibly even for eternity.

Although astrophysics is integral to modern cosmology, most theories place far more emphasis on the age-old astronomical traditions of geometrical and mechanistic coherence. The emphasis is on the grand design rather than the detailed mechanisms. Accordingly, Einstein, whose rationalist attitude informed modern cosmology, dismissed the importance of astrophysics, which arose from the new optical technology of spectroscopy and dominated late nineteenth-century astronomy, when he said, "I want to know how God created this world. I am not interested in this or that phenomenon, in the spectrum of this or that element. I want to know His thoughts, the rest are details."[1] This search for perfection was taken one step further by Hermann Bondi and Thomas Gold when they proposed the steady-state theory in the 1940s, postulating a Universe boundless in both space and time. By contrast, Fred Hoyle saw the steady state primarily as an astrophysical theory where chemistry and physics take precedence over understanding the geometry of space-time. Hoyle intimates that God's thoughts *are* what Einstein terms the "details," and not just the broad structures. Like the fashion designer Jean Muir, who maintained that "God lurks in the details," Hoyle might claim that God lies in the subatomic levels rather than in the grand design. He believes that it was this tradition of attention to empirically testable physical detail, originating in the astronomy of the nineteenth century, which gave the steady-state model its legendary elegance and beauty, comparable to the "exquisite subtlety" of the rest of science. This contrasts with the commonly held view that the steady state owes its beauty to the elegance of its conception, its problem being that it fails to correspond to reality. It is also generally believed that the steady state contains no more testable physics than the big bang does. Hoyle is, in effect, the pot calling the kettle black.

Both the big bang and the steady state are essentially realist, or rationalist, theories, though both were made possible by physics developed in the nineteenth century. Astrophysics enabled cosmology to move beyond the traditional philosophical and theological speculations that had characterized it until this century. Nevertheless, neither theory has abandoned the ideological motivations, or the traditional desire to create a coherent and imaginative view of the Universe. Nor can they do so. Cosmology is a uniquely inaccessible scientific discipline that cannot appeal directly to repeatable circumstances or laboratory tests in order to establish the validity of its arguments. To a large extent, any cosmology has to depend on the rational coherence of its predictions, and its acceptability therefore depends heavily on judgment rather than objective fact. Although it can appeal to testable physics, and can even suggest new physical laws that may or may not be testable, the models of cosmology have to be essentially rationalist in conception. Accordingly, because much of it is empirically inaccessible, many of the ideas of twentieth-century cosmology are almost impossible to falsify conclusively. Although there is some truth to Hoyle's belief that the creation mechanism of the steady state is more empirically accessible than that of the big bang, the significant difference between the two theories is really philosophical and aesthetic. Both the steady state and the big bang are based on Einstein's theory of general relativity. However, the big bang follows the Judeo-Christian tradition of a Universe created out of nothing at a finite time in the past, whereas the steady state recalls the idea of a Platonic Universe, eternal and perfect. This perfect cosmological principle contains the idea that the Universe is isotropic and homogeneous, that it is broadly the same throughout space and time and that it looks the same today as it did billions of years ago.

Like Newton's laws, Einstein's theory of gravity was devised in the context of the ancient belief in a static Universe. However, Einstein realized that something like antigravity was needed to prevent the Universe collapsing. Gravity holds the Universe together, but what keeps things apart? Until the late 1920s, he could offer no ontological solution since his theory of general relativity contained nothing resembling antigravity. He made the Universe balance by using an ad hoc parameter in his equations which he called the cosmological constant, which he felt would do service for the as yet inexplicable physical force keeping things apart. He later called this his biggest mistake. He had missed a great opportunity to make the stunning prediction that the Universe is expanding. Also, it soon became evident that a static system which relies on a cosmological constant to counteract the effect of gravity is inherently too unstable to serve as a sustainable model for the Universe. It has to be dynamic to survive. With the wisdom of hindsight, Einstein saw that he should not have stuck to his preconception that the Universe is static, but accepted what his mathematics was trying to tell him. There was no need for him to impose a cosmological constant since the Universe really is expanding, which explains why gravity had not yet won the day and obliterated the Universe. The physics that keeps the Universe in some sort of balance is the well understood phenomenon of an explosion of energy overcoming the pull of gravity.

Thus the early big-bang model seemed to offer the solution to the problem of how the Universe can exist in anything like the form we observe. As a broad hypothesis it seemed to do the trick. However, the devil as always lurks in the details or, as Hoyle would contend, in the inability to produce details. It soon became apparent that the big-bang model had a similar weakness to Einstein's static universe. The initial conditions of creation would have to be extremely finely balanced—otherwise the Universe would either have collapsed shortly after its creation, or it would long ago have expanded into an infinitely attenuated vastness, well beyond the gravitational influence of its constituent parts. Either gravity or creation's energy would have been vanquished, and the other left to reign alone.

Also, in the 1930s it was not clear that the expansionary force of creation was enough to overcome the force of gravity. There might still be the need for a cosmological constant to explain how a Universe as young as the contemporary estimates of 2 billion years could contain as much matter as was observed. In other words, in order to account for the fact that the Universe was expanding at the high rate associated with a relatively young age, there had to be either far less matter, and therefore gravity slowing down the expansion, or there had to be some form of antigravity other than the energy of creation. As independent measurements increased the age of the Universe, so more matter was allowed and with it the decreasing need for a cosmological constant. Although astronomers subsequently lost interest in it, the need for a cosmological constant was not entirely removed until Alan Guth came up with the theory of inflation in 1981.[2] This theory, which we shall look at in detail later in the book, not only provided a creation mechanism which explained why the Universe was in balance, but also implied that the Universe was in a critical state between infinite expansion and eventual collapse, and contained vastly more matter than was observed.

The steady state solved the problem of accounting for what stops the Universe collapsing with what Hoyle claims is a detailed, testable mechanism well within the bounds of empirical physics. However, it implicitly conflicts with the second law of thermodynamics, which requires the overall increase of a quantity known as entropy, essentially a precise mathematical formulation of the concept of chaos. In other words, as a closed system gets older, it degenerates from a state of order into featureless, unstructured chaos. To my mind, the application of the second law of thermodynamics to the expanding Universe as a whole, which is not a closed system, is an example of a deeply embedded scientific prejudice rooted in Judeo-Christian creation mythology, and the attendant idea of the linearity of time with a beginning and an end. Thus it is no accident that the big bang holds on to the traditional idea of the inviolacy of the second law of thermodynamics and the increase of entropy. This essentially one-way process can be seen as a way of defining the direction of time, as is evident from Stephen Hawking's somewhat convoluted account of the "arrow of time" in *A Brief History of Time*. The notion underlying standard cosmology is that the creation of all matter and so all energy represents an initial state of supreme order which gradually dissipates into the dark, cold chaos of an energy-dead Universe. Nature simply does not have it within itself to create or recreate the bulk of its own energy and matter. Similarly, many people believe that life cannot arise by itself, but needs an outside agent to get it going, hence the skepticism that still surrounds Darwinian evolution. It seems to me that equating the two very different notions of time and change of entropy is at the root of the problem.

Hoyle sees all this as fundamentally incoherent, involving either infinite regress or an unimaginative retreat into naive metaphysics using the asymmetrical law of entropy, which prescribes a one-way journey into chaos. This results in the idea of the Universe as unselfmotivated and dying of old age and neglect. Hoyle's picture of the Universe is closer to the idea of an eternally self-generating, living entity that is not merely a passive victim of physical laws. The Universe is absolutely everything, including the laws of physics. These laws do not have any kind of independent existence. Hoyle feels that it makes no sense to separate the Universe from any of its aspects whether it be the counterentropic phenomena of life and man's intellect, or the concept of time itself. If time is defined by the second law of thermodynamics, then what does life do to time? Is time reversed

31

every time a baby is conceived? Hoyle believes it is intellectual nonsense to propose a beginning or end to time, just as it is nonsense to talk about a beginning or end to space.

In terms of observational evidence, there was not much to choose between the early big-bang and steady-state theories. Also, since they were both committed to some sort of creation model, they both shared the same, almost intractable weakness of having to deal with the metaphysical notion of creation out of nothing. However, since the Universe has always been in existence, the steady-state model had the advantage of not having to propose the creation of absolutely everything. Nevertheless, both theories had either to remain vague about the idea of creation out of nothing, or appeal to the mysterious working of a divine Creator. However, the steady-state constant creation model accounted for the fact that the Universe has not collapsed upon itself and, indeed, appears to be expanding, without having to invoke a special set of initial circumstances. Since the output of creative energy is continuous, the force of gravity is always overcome by the expansionary pressure of creation. Thus the steady state neatly circumvented the problem of explaining how things could be kept in perfect balance in a static Universe, or how the initial conditions of the big-bang creation could be so critically balanced as to sustain any sort of dynamic equilibrium over time.

However, the current consensus is that it is the steady-state theory which has been refuted. In its original form this theory was confronted with three major observational challenges. The first was to explain how the number of astronomical objects that emit radio waves, or "radio sources," could have been so much greater in the past, thus violating the perfect cosmological principle of a steady-state Universe. As we shall see, this was largely a spurious problem created by Martin Ryle's overenthusiastic attempts to discredit Hoyle. This argument became much less conclusive when the radio source counts were done properly. The second challenge was to account for the production of light elements such as helium and lithium in their currently observed abundances, which are too great for these elements to have been synthesized within bodies such as stars. Proponents of the big bang argued that the correct abundances of the light elements were a natural consequence of the theory, and though the details have remained somewhat controversial, this has on the whole been regarded

as a successful "prediction." At first it was not clear how light elements could be produced in the steady-state model, but Hoyle has
since incorporated a mechanism that gives abundances in agreement
with observations, although his solution has been criticized for being
contrived. Rather than the constant creation of matter and energy
within stars, Hoyle's revised model is of a sort of oscillating bubble
Universe. Each oscillation within our bubble takes hundreds of billions of years. The expansionary phase is initiated by the creation of
matter and energy under the extreme gravitational pressure of the
collapsing bubble, or a "mini big-bang," which is both caused by and
overcomes the force of gravity. Eventually gravity reins in this expansion and pulls everything together again into a grand collapse, thus
providing the conditions for triggering another explosion of creation,
and so on. So instead of an antigravity, constant creation mechanism
which, like Atlas, eternally holds up the Universe, the revised steady-
state model portrays it as a seesaw mechanism of creation and destruction in which the ascendancy of either gravity or creation's energy leads to its own downfall and the ascendancy of the other. Here
the Universe, or at least our bubble in the Universe, is like the phoenix,
eternally reborn from the ashes of its own destruction, the one always
giving rise to the other. The whole Universe is infinite and on average
expands indefinitely, while in each bubble, as with other local time
such as days, months, and so on, time is cyclical.[3] Everything comes
round again.

The deciding factor in the controversy between the big-bang and
steady-state theories was generally seen to be the discovery in 1965 of
a faint uniform glow of microwave radiation apparently pervading
the whole Universe. This was later called the cosmic microwave background[4] and possessed three remarkable properties. First, it was extremely uniform over the whole sky, varying, as was later found, by
less than one part in a million. Second, the shape of the energy
distribution was precisely that of a "black body" heat source; and
third, the temperature was 2.7 degrees above absolute zero. By an
astonishing coincidence, at the same time as the microwave background was discovered, Richard Dicke and others published a paper
predicting the first two of these properties.[5] Such a dramatic prediction has always counted very strongly in a scientific theory's favor, and
from then on the steady state, which had made no prediction about a

microwave background, was no longer seen as a serious contender to the big bang. Although, subsequently, the microwave background was given an explanation within the framework of the steady state, by invoking a population of needlelike iron molecules that absorb ambient starlight and re-radiate it in a way that mimics blackbody radiation, it is hard not to see this as a contrived addition to the theory, with none of the elegance of the big-bang explanation. The only serious counterattack on the big bang was to point out that the theory did not predict the temperature of the microwave background, which depends on the value of other parameters, especially the age of the Universe. This point was sometimes glossed over when arguing the case for the big bang.

Big-bang dissidents[6] take the view that the big bang has been refuted by observational evidence and should therefore, according to the precepts of Karl Popper, be abandoned. Instead of abandoning the theory, big-bang supporters have introduced a number of ad hoc additional assumptions with no observational support, and with the sole purpose of reconciling the consequences of the big bang with observation. Among the predictions that the dissidents claim have been falsified, two are of particular interest. First, they point out that the big bang predicts a Universe that has not had time to create the large structures into which galaxies are organized. They claim that the various ways in which this problem has been circumvented are all contrived, and in any case tend to create other difficulties. For example, one can invoke a cosmological constant that will have the effect of increasing the age of the Universe to large values. However, too large a value would raise the numbers of quasars and galaxies with high redshifts to beyond the observed levels. Alternatively, one can postulate that there is a difference between the distribution of visible matter, which reveals the clumpy structures we see, and a more uniformly spread dark matter representing the bulk of the Universe, which has not had time to collapse into structures. The difference between these two types of distribution is described by a bias factor. The dissidents point out that, apart from being a completely ad hoc solution, there is no obvious mechanism by which such a segregation could occur to a sufficient degree.

The second prediction of the big bang which they claim has been contradicted is that the microwave background should be inhomogeneous, in the sense that when we measure the microwave flux in opposite directions we are looking at two parts of the Universe that could have had no contact with each other since they were isolated by different "event horizons" at the time the background radiation was emitted and so should be at unrelated temperatures. In other words, at the time that matter decoupled from light, giving rise to what is now observed as the cosmic background radiation, the finite speed of light and transmission of information made it impossible for different parts of the Universe to "know" or affect one another. So how is it that when we observe the microwave background radiation we see such an extremely uniform temperature? The solution to this problem, with which most astronomers are now very comfortable, is the theory of inflation conceived by Alan Guth. In this picture, which is described in greater detail in Chapter 6, the entire Universe that we now observe was once a minute volume in thermal equilibrium. This cosmic speck expanded by an almost unimaginable amount to form the Universe that we now see, maintaining the uniformity of its temperature as it cooled. Some dissidents claim that, however elegant this picture might be, it makes no observationally confirmed prediction, and should be regarded as an unnatural way of solving the "horizon problem."

In the event, although some nagging problems refused to go away, the big-bang theory carried the day, and has given its name to our current model of the Universe. Nevertheless, inflation, which had so elegantly solved the horizon problem in the big-bang theory, has much in common with some of the ideas of the steady state. The chapters that follow trace the subsequent development of the big-bang theory in the standard model, and with the benefit of hindsight show that current ideas about the large-scale structure of the Universe blur the old distinctions between steady-state and big-bang cosmologies and make it clear that in the classic battle between Hoyle and Ryle, neither was right or totally wrong.

Hoyle's work, and especially his theory of nucleosynthesis, is undoubtedly Nobel Prize–winning material. He was one of the first to apply nuclear physics and Einstein's theory of relativity to cosmology, and he pioneered work on the ages and temperatures of stars which is crucial for estimating the age of the Universe and therefore for shaping our cosmological models. He also claims credit for predicting the existence of quasars,[7] though it seems to me that this is an overinterpretation of Hoyle's description of processes and events in the extreme gravitational conditions of very large and dense compact bodies. However, in 1983 the award for Hoyle's considerable achievements was denied him and given instead to his collaborator, William Fowler. This was considered by many to be an enormous injustice, which could only be explained by the rumor that the Cambridge astronomical establishment had advised the Nobel committee against awarding the biggest prize in science to a maverick like Fred Hoyle. Fowler, on the other hand, had the full support of his Californian establishment.

Hoyle's remarkable achievements were seen to be canceled out by the death blow to the early version of the steady state. The two Bell telephone engineers, Arno Penzias and Robert Wilson, who accidentally detected the cosmic microwave background shared the 1978 Nobel Prize for physics. Despite his eccentric resistance to, and disdain for, the whole business of awards, and despite his erroneous claim that his Cavendish Laboratory would produce the evidence to unseat Hoyle's ideas, Martin Ryle eventually had the Nobel Prize for physics thrust upon him in 1974 for his role in developing what was then considered to be the cutting edge of observational technology. Radio astronomy was regarded as the means by which the question of the correctness of the big-bang theory would be settled, and it was seen as the future of cosmology. Although Ryle had set himself against Hoyle and repeatedly stated that his Cavendish Laboratory had falsified Hoyle's ideas, these claims always turned out to be untrue, and radio astronomy never quite fulfilled those high expectations. But Ryle's firm stance against the steady state to some extent vindicated him, and also contributed to his stature as a champion of big-bang cosmology.

So Ryle, radio astronomy, and the big bang were together seen as victorious. Ryle was lionized as the beleaguered guardian of truth who, bearing the bright torch of observational evidence, eventually triumphed over the dark falsehoods of Cambridge's theoreticians. Hoyle, along with the steady-state theory, was discredited, humiliated, and, rather like the hate figure in George Orwell's *Nineteen Eighty-Four,* held up to ridicule and antipathy as the enemy of scientific truth.

Well before the discovery of the microwave background, Ryle organized a press conference that, unknown to Hoyle, was specifically designed as a public condemnation of the steady state. Ryle claimed that he had conclusive evidence against Hoyle, who was unprepared for such denunciation when he accepted Ryle's invitation to attend. Hoyle was seated alone on the stage from where he had to fend off hostile questions from excited spectators.[8] Ryle's Cavendish Laboratory had done a count of radio sources, which he maintained were more numerous at large distances, from which it could be inferred that they were more common in the past. This could be accommodated within the evolving universe of the big bang but was inconsistent with the homogeneity of the steady-state model, where radio source numbers should be the same at all times and in all places. Although Ryle's so-called damning evidence soon turned out to be wrong when the counts were done properly, there was no public retraction, and the impression remains to this day that Hoyle was spectacularly misguided.

What dreadful evil was the irrepressible Fred Hoyle guilty of? Was it his refusal to meekly accept the validity of any evidence apparently contrary to the steady state, or was it, as Hoyle contends, mainly his unforgivable crime of belittling the achievements of Ryle's Cavendish Laboratory?

According to Hoyle, Ryle was bent on revenge for the humiliation he had suffered in 1951 when he was proved wrong in his flat denunciation of Hoyle and Thomas Gold for suggesting that, contrary to Ryle's belief, most radio sources were very distant. Hoyle maintains that

> Ryle's motivation in developing a program of counting radio sources, a program that was to occupy a major fraction of his group over the next ten years, was to exact revenge for his humiliation over the radio star affair. This was to be done by knocking out the new form of cosmology with which Gold, Bondi and [Hoyle] were associated.[9]

Although this might explain the manner of his excommunication from the astronomical community, which looked more like the result of a witch hunt than of a scientific argument, surely the reasons for the ascendancy of big-bang cosmology cannot be entirely attributed to the vindictiveness of an influential rival.

There must be more objective reasons for the general acceptability of a scientific idea than the primitive herd instinct of following the victor of a personal power struggle. The reasons are undoubtedly more complicated, especially where large financial resources and national or institutional prestige are at stake. Although it might boil down to herd instinct, the need to keep one's job or to provide secure employment for a large number of scientists is also a factor. Ryle must have felt that Hoyle was threatening his rice bowl and that of everyone associated with the Cavendish. Those associated with Hoyle must have felt like rats on the sinking steady-state ship. Either they had to join the big-bang fleet, or go down with their redoubtable captain.

Although this type of social mechanism is very common, its operation is seldom recognized in science, especially by those who have an exaggerated regard for scientific objectivity. Steven Weinberg and Karl Popper, for instance, were enraged by the suggestion that science is a social process beset by vested interests and personal power struggles, and that scientific "truth" is therefore determined by consensus, or even complicity.[10] But to my mind this self-perpetuating process is the only explanation for a number of widely held scientific beliefs, most notably the values of the Hubble constant. Measurement of this parameter, which specifies the scale of the Universe, has been the preserve of powerful observational astronomers such as Edwin Hubble, Gerard De Vaucouleurs, Allan Sandage, and, recently, those using the Hubble Space Telescope. To my mind, their methodology for measuring the Hubble constant has fundamental limitations, but such is the authority and the weight of vested interests surrounding the search for this important cosmic parameter that it has recently been accepted largely without question, and consequently exerts a powerful but confusing influence on the course of modern cosmology.

However, before going into the whole business of observational astronomers bickering over the value of the Hubble constant and the consequences of the Hubble Space Telescope's influential publicity machine, I would like to reflect on and examine some of the philosophical issues such as the relationship between theory and evidence that underlies not only the big-bang/steady-state controversy but science in general.

4

REALITY IN THE DARK

It is not *how* things are in the world that is mystical, but *that* it exists. We feel that even when all possible scientific questions have been answered, the problems of life (or existence) remain completely untouched. Of course there are then no questions left, and this itself is the answer. The solution of the riddle of life (or existence) in space & time lies *outside* space & time. (It is certainly not the solution of any problems of natural science that is required.) What we cannot speak about we must pass over in silence.

—LUDWIG WITTGENSTEIN, *Tractatus Logico-Philosophicus*

WE HAVE SEEN THAT A MAJOR REASON for Fred Hoyle's attack on the big bang is his conviction that it is at heart a metaphysical theory and therefore hardly any more scientific than, say, Buddhism. He sees himself as an empiricist and the steady state as an empirically accessible and hence scientific theory as opposed to being fundamentally rationalist and therefore, like religion, dependent on unscientific factors such as personal conviction. What is the difference between rationalism and empiricism and what makes the requirement for objective evidence so important? Why should we not just think up a good idea and accept it as the truth rather than go to all the expense and effort of finding evidence? Surely all we need to do is exercise our innate ability for recognizing the truth and determining reality, or alternatively, we could perhaps simply consult those with superior intellectual or psychic powers for apprehending reality without the need for evidence.

Obviously this is a philosophical issue. Science is and always has been a branch of philosophy, despite the fact that many prominent scientists feel that it has outgrown the restrictive bondage of its parent and therefore that it no longer resembles philosophy. Others believe that philosophy no longer has anything to offer simply because it has abdicated its responsibility to science. Nevertheless, almost all science writers have something to say about underlying philosophical issues, even if it is just to complain about their destructive influence on scientific methodology. For instance, Steven Weinberg believes that the Vienna School of logical positivists,[1] which was led by one of the greatest philosophers of our time, Ludwig Wittgenstein, had a detrimental effect on the early development of particle physics.[2] This view was shared by Albert Einstein, and caused him to defect from the field of quantum mechanics and to formulate his highly rationalist theory of general relativity. Einstein repudiated the logical positivist idea that the route to truth is via an inductivist assemblage of data. He believed that good scientific theories are mainly the product of the intellect rather than of actual sensory experience.

The disagreement between Einstein and the other founders of quantum mechanics hinges on an age-old dichotomy, first formulated by Aristotle and Plato, between the roles of the intellect and experience in determining reality. This dichotomy continues to provide the basis for nearly all philosophical issues, and is still the main source of ideological disagreement between scientists or at least between different branches of science. For instance, the theory of Darwinian evolution as expressed by Richard Dawkins[3] is in head-on conflict with the Platonism of "creationist" interpretations of evolution which is usually espoused by nonbiologists such as the great astrophysicist Fred Hoyle[4] or the theologian Hugh Montefiore.[5]

This type of conflict exists because of a fundamental difference of opinion as to how we can know whether or not our ideas reflect objective reality. Plato gave no weight to evidence since he believed that the only way we can apprehend essential reality is through our imagination and intellect. He maintained that ultimate reality represents a level of perfection and order which we can only imagine since the material world as we experience it is chaotic, imperfect, and uncertain.[6] Thus Platonism is synonymous with realism, idealism, and rationalism. Plato felt that the abstract language of mathematics represents the structure of reality. Aristotle, on the other hand, maintained that we build up a picture of reality simply from observation, experience, and other evidence.[7] This inductive process is the basis of the empiricism that many people believe should represent the underlying philosophy of scientific methodology.

Wittgenstein and Dawkins feel that Platonism leads to a whole lot of metaphysical guff, whereas Weinberg and other intellectual Brahmins such as Stephen Hawking see Aristotelian methodology as not only unimaginative and stifling, but also responsible for scientific nonsense such as the indeterminist epistemology of quantum mechanics. Einstein felt that, because of the influence of Wittgenstein and others, particle physics was too rigorous in its application of the Aristotelian principle of building up a picture of reality by an accumulation of evidence. It had therefore lost itself in an intellectual wilderness where it could not see the forest for the trees, and consequently was coming up with all sorts of crazy ideas. But this approach has been remarkably successful, and has amazing predictive power, though nobody can understand why. It actually works in practice, though thinking about it too much can drive intellectual realists insane. Crazy or not, quantum physics is the foundation of most of the greatest technological developments of our age. Perhaps this is the source of the stock figure of the scientist as a deranged genius.

Weinberg is more concerned about the obstructive restriction of excessive reliance on observational evidence. He has this to say:

> Despite ... the defection of Einstein, the theme of positivism has continued to be heard from time to time in the physics of the twentieth century. The positivist concentration on observables like particle positions and momenta has stood in the way of a "realist" interpretation of quantum mechanics, in which the wave function is the representation of physical reality.[8]

Weinberg is both a particle physicist and, like Einstein, a mathematical physicist. Like most cosmologists he is a Platonic realist or rationalist, and consequently believes that particle physics is successful despite being founded on the inductive empiricism of positivism. Weinberg's argument seems to rely on the by no means self-evident assumptions that only "realist" interpretations of scientific models are useful, and that unobservables such as wave functions are somehow more real, probably because they can be elegantly expressed in mathematical terms. However, it can be argued that it is because of such realist or determinist tendencies within quantum mechanics that this branch of science has ended up philosophically confused and hence counterintuitive. Perhaps it is because quantum mechanics tries to reconcile empirically derived indeterminism with the realist determinism of classical physics that it often seems irrational and contradictory. Maybe if quantum physicists had stuck to their guns and resisted the undermining influence of Einstein and his adherents on their early positivist rigor, they would have developed a radically new physics that makes absolute sense and would therefore appeal to our intuitions in the way that Darwinism does.

In his attack on positivism, Weinberg glosses over the extent to which this philosophy is responsible for our popular idea of what distinguishes science from, say, religion.[9] This idea of science can be summed up in Brian Malpass's words: "In modern science, experiment and observation are king. A law or theory, no matter how ingeniously fashioned, no matter how elegantly phrased, is like the present day football manager—only as good as the last result."[10] But the fact is that Platonism has permeated every fiber of Western culture. For example, Christianity is a form of Platonic realism. Thus it is no accident that a rationalist theoretician like the cosmologist Stephen Hawking, can, without undermining his own scientific position, maintain that the material Universe represents the mind of God.[11] On the other hand, the empiricist Charles Darwin, who came up with his idea of evolution through acute and meticulous observation of the natural world, could not encompass the idea of God within his theory without completely negating its explanatory power. The principle of natural selection and the idea of God are mutually exclusive. As Dawkins points out, there can be no compromise: either Darwinian evolution is true, or God the creator exists, but not both.[12]

Down through the ages, and to a large extent in our time, empiricism has been regarded as the devil's work in sowing chaos, confusion, and doubt. Galileo[13] committed a crime against certainty and the natural order by trying to explain the Universe in terms of what he could observe. Darwin's ideas are not only correctly seen as the enemy of religion, but also as the cause of wicked unnatural practices such as genetic engineering. Physicists are perceived as boring and abstruse, and also the cause of most of our modern evils from nuclear weapons to the anarchic computer Internet, unless, of course, the physicist happens to be a theoretician and a Platonic realist. Then he is revered as a sort of high priest who, like Plato, probes the dark mysteries of the Universe through the power of his intellect. Thus Brian Malpass alleges that Stephen Hawking, "the supreme cosmologist, is said to have looked through a telescope only once in his life. He got such a headache that he had to have a lie down to get over the experience."[14]

Because our culture is imbued with Platonism, it dominates our way of thinking. So although we pay lip service to the value of evidence or observations and, in accordance with Wittgenstein's dicta, piously assert that scientific theories, like religious ideas, cannot in themselves tell us anything about reality, we nevertheless believe, act on, and are even spellbound by good theories, regardless of whether or not there is any supporting evidence. Accordingly, Sir Arthur Eddington, who is famous for unsuccessfully trying to confirm and explain Einstein's theories of relativity, once remarked that observations should not be believed unless they are supported by theory.[15] It is not unusual for astrophysicists to construct entire cosmologies with very little reference to observational data.

Even our most influential philosopher of science, Karl Popper, models his idea of the scientific process along Platonic lines. According to him, intellectually derived theory is king. Observational data and experimental evidence are merely unruly vassals, always threatening to topple their sovereign. A theory is dreamed up and then judged scientific if it is testable by experiment and observation. But no matter how many times it is confirmed, there is always the possibility that something will come along to falsify this theory. Therefore no scientific theory can ever represent truth. So for instance, the proposition that the Sun will always rise in the morning can never be regarded as true, merely scientific. All statements about matters of fact and existence are only provisionally true, and it is only a matter of time and effort before their falsity is demonstrated. In Popper's model of science the role of evidence is not to establish the truth, but merely to provide temporary support for sustaining the reign of an intellectually derived idea or to furnish ammunition for eventually demolishing it.[16]

This view of the scientific process is, with some minor variations, considered the standard. Consequently it is used both by those who

43

consider themselves exponents of pure science, and by those who are opposed to some aspects of science and wish to undermine the validity of inductive disciplines such as sociology and Darwinian evolution. Sociology, anthropology, archaeology, and so on, are often dismissed by physicists as "pseudosciences" because their ideas of the world are painstakingly and imperfectly built up from an accumulation of evidence, rather than from grand intellectual ideas which represent the kind of order and coherence that cannot be achieved by the inductive method.[17]

Because Darwin held evidence to be king, and built up his picture of reality from what he could actually observe, his theory of evolution is sometimes regarded as unscientific. Consequently the principle of natural selection is regularly accused of being untestable or unfalsifiable by scientists and theologians alike. But how can evidence falsify itself? Well, of course it can't. Evidence can only falsify theories. So this leads to a further criticism of Darwin's idea: that it is not a proper scientific theory, but at best a comprehensive catalogue of natural phenomena which has very limited explanatory power. It brings us no closer to the great imponderables such as how life started in the first place, or what the grand design and purpose of it all is. These Platonic questions represent the underlying philosophy of so-called big science and mysticism alike. Both uphold the principle of establishing natural "laws" or "necessity." One could argue that theoreticians have more in common with theologians than with experimental or observational scientists, who are seen as mere workers subservient to the great intellectual taskmasters engaged in the activity of pure science. Consequently, these sacred and scientific Platonic realists often join forces against Aristotelians such as Charles Darwin or his modern disciple Richard Dawkins, who are in consequence, astonishingly, still regarded as controversial in spite of, or perhaps because of, the overwhelming weight of evidence in their favor.

Darwinism is the complete opposite of idealism or of any a priori preconceptions that characterize the ideas of Platonic realism. It does not have built into it the assumptions that there is a grand design, that there are necessary laws, or that models of reality must implicitly point toward some ultimate unifying explanation for everything. Because Darwinism starts from the evidence, it bypasses the approved stages in the construction of scientific ideas such as mathematical argument or appeal to established authority. For instance, in Darwin's time the age of the Universe was derived from the authority of the Bible. Even great scientists like Johannes Kepler and Isaac Newton relied on the catalogue of biblical "begats" when calculating the age of everything in creation. By contrast, Darwin adopted the maverick hypothesis proposed by empirical geologists that the Earth is enormously old, and that the Bible is a completely unreliable authority for dating the Universe. However, Darwinism does not even ask questions like, "Is this likely or credible?," or "Does this accord with established ideas?" It simply points out that this is the way things happen to be, that here is the evidence, and challenges anybody to find a single counterexample or a better explanation. Apart from the fact that one cannot find any such falsifying evidence, it is also impossible to refute because it is not an argument. As Dawkins points out, there is not even the beginnings of an argument against Darwinian evolution or a better interpretation of the data, apart from variations on the creationist theme which amount to an appeal to the authority of long-established Platonic idealizations.[18]

The immense authority of Popper's prescriptions of what constitutes science is used to lend a spurious plausibility to such creationist challenges or, as Dawkins calls it, "[fatuous] argument from personal incredulity."[19] Regardless of the fact that Darwinism owes its existence to a huge body of tangible evidence, the realization that it cannot be falsified or refuted is still taken as a reason for giving it the same status as the claims of mysticism. Hence theologians feel entitled to impose on Darwinism all the usual arguments for the existence of a Creator, such as the "argument from design." This is the attempt to prove the existence of God by pointing out all the wonderfully clever things that exist, and concluding that there must have been an intelligent designer since such beauty and complexity could not have come about by mere chance.

Physicists, who, like mystics, have an abhorrence of anything that suggests a lawless Universe ruled by chaos or chance, use variations on this argument from design to undermine the validity of the principle of natural selection. For instance, Fred Hoyle claims that life cannot be explained simply in Darwinian terms since it is statistically untenable for life to have originated on Earth by mere chance, let alone developed in all its magnificent complexity merely by hit and miss. There must be some preexisting master plan directing the path of evolution. Hoyle suggests that the information, the master DNA, analogous to computer software, which contains all possible variations of life, came to us from outer space.[20] This is a mathematical argument which depends on Hoyle's belief that the Universe is infinite, that it had no beginning and will have no end.

However, Dawkins maintains that it is invalid to subject the theory of evolution to such mathematical or Platonic realist analysis since it is not part of its terms of reference. Evolution is not a game of chance; quite the contrary. It is not a theory about what might be, but an explanation of what is manifestly the case. One simply cannot make statistical analysis part of the explanation of what already exists. It is the other way round. What is already the case is used to predict what might happen. Once something happens it is no longer unlikely or likely, it just *is*. Statistics is just a complicated method of counting. There is nothing lawlike about it. It does not describe reality, it merely quantifies it. When we go about looking at the world we are never astonished that what we observe is statistically unlikely. We might be astonished to find ourselves with a winning lottery ticket, but statistical unlikelihood is part of the game of gambling. So subjecting the principle of natural selection to this type of analysis is simply an invalid attempt to quantify personal incredulity. It is a disguised form of the creationist argument, which itself is an equally invalid attempt to impose Platonic terms of reference which are not an intrinsic part of the theory of evolution. Darwinism is not a treatise about the existence or nonexistence of a Creator, nor is it a mathematical argument. It is not a type of abstract metaphysical discussion about the guiding principles of the Universe, no matter how much theologians and other Platonic theoreticians try to crush its neutrality and force it into this mold because they feel threatened by the total godlessness and complete absence of ultimate purpose implicit in the self-sufficient principle of natural selection.

The great eighteenth-century Scottish empiricist David Hume pointed out that we ordinarily make sense of the world by inductive reasoning.[21] This is not strictly speaking logical. Inductive or commonsense thinking is not the abstract intellectual process that is usually encapsulated in formal logic or mathematics. Such Platonic reasoning is of an order different from the complicated way we actually go about making sense of the world. Hume maintained that such natural reasoning, the ability to see the connection between things, which relies on the irrational concept of causality, although illogical, is, "to us, the cement of the Universe."[22] Without the animal instinct to think inductively or causally, we would be unable to make sense of anything whatsoever. The Universe would be no more than a set of disconnected sense impressions. We would, in effect, be incoherent occupants of a realm of chaotic contingency.

Following in the tradition of Hume, Wittgenstein maintained that not only is inductive reasoning more interesting and more important for our understanding of the objective world, but formal logic and mathematics are just metalanguages or languages about languages. They are pseudolanguages and as such tell us nothing about the real world since such metalanguages represent no more than our way of thinking about our way of thinking. Metalanguages are self-referential and hence viciously circular. They are systems of tautologies. They cannot get beyond themselves. Therefore we cannot get at the truth about the way things are through formal or mathematical logic, any more than we can transcend the self-sufficient claims of mysticism or make any inferences about the world from the rules of chess. The models of religion or theoretical physics, for example, can in themselves tell us nothing about reality, no matter how coherent or elegant or philosophically satisfying they are. Accordingly, Timothy Ferris maintains that truth is beautiful, but what is beautiful is not necessarily true.[23] But it seems that what is true may not even be beautiful. In other words, contrary to Weinberg's belief,[24] beauty is not even a criterion for truth. Certainly Darwin's ideas fail to meet the standard of beauty attained by the theories of great cosmologists such as Newton or Einstein. Bertrand Russell's colleague, the mathematician Alfred Whitehead, declared that "Darwin is truly great, but he is the dullest great man I can think of,"[25] and Darwin wrote in his autobiography that "my mind seems to have become a kind of machine for grinding general laws out of a large collection of facts."[26] Where is the elegance and beauty in all that? Timothy Ferris claims that "*The Origin of Species by Means of Natural Selection* is objective to the point of bloodlessness." Nevertheless, he goes on to say that "indeed the book was so detailed and modest that it struck many readers as self-evident."[27]

47

This is the strength of empirical induction. It may be dull and clumsy, but it is the way we ordinarily make sense of the world and it therefore appeals directly to our intuitions. It is not exciting because it does not stretch our credulity or require us to suspend our natural "commonsense" way of thinking. However, both Wittgenstein and Dawkins take great pains to point out the quiet beauty and deceptive complexity inherent in ordinary empirical ideas.

Like Hume, Wittgenstein elevates the analysis of inductive reasoning to a great philosophical pursuit as the only way of really making sense of the objective world. Dawkins sees the simplicity and obvious truth of Darwin's principle of natural selection as intensely beautiful. So, it seems that even criteria for scientific beauty are debatable; furthermore, truth may be ugly and messy. As Dennis Sciama said when the weight of observational evidence eventually forced him to abandon the beautiful, all encompassing steady-state cosmological theory in favor of the crudity of the big bang: "The Universe is in fact a botched job, but I suppose we shall have to make the best of it."[28]

Wittgenstein's famous, or infamous, statement that the only task left to philosophers is the analysis of language is considered by Stephen Hawking, among others, as an enormous cop-out. Hawking felt that Wittgenstein was saying that philosophy can no longer handle larger issues like ontology and had to retreat into the trivial business of linguistics.[29] But natural language mirrors the way we actually make sense of the world, so in terms of understanding reality it is a better vehicle than mathematics or formal logic. Also, natural language encapsulates far more complexity than the artificial languages of mathematics and formal logic. Thus the analysis of language is a much larger intellectual activity than, say, the study of mathematics, which is just a small subset of linguistics. By these criteria, physics is pretty low down in the order of intellectual activities.

Much of physics is philosophically naïve, largely because of the assumption that only logically transparent artificial languages like mathematics can represent the higher intellect and therefore provide a route to higher truths. Wittgenstein came to realize that this traditional and time-honored belief is a fallacy leading philosophers into all sorts of conceptual nonsense, and that our far more complex natural reasoning is the only route to reality. For this reason he decided that the analysis of the opaque structure of natural language is the task of philosophy. He most certainly did not intend his statement to be construed as an invitation to physicists to take up the cudgels for what he considered the discredited methodology of traditional rationalist philosophy.

The extent to which the modern sciences can be differentiated from mysticism or traditional philosophy is the extent to which such systems are also natural inductive systems. Thus the ideas of cosmology are often difficult to distinguish from those of metaphysics, especially when we just cannot imagine how to go about finding supporting evidence. Like philosophical or religious ideas, such models are usually challenged only on the basis of their mathematical or logical coherence, and on the extent to which they contribute to or conflict with the existing framework of accepted ideas. But theories that are in principle easy to refute or support with empirical evidence are thought of as more characteristically scientific.

There is a great deal of substance to Wittgenstein's idea that we are often led into incoherence through the fallacy that formal logic and mathematics somehow represent the natural order of the Universe. But the most common result of this fallacy is the belief that what is unassailable in mathematical physics must represent immutable truths about nature. This leads many physicists into the illusion that they can access essential reality directly by mathematical reasoning alone and that there are some universal truths about which we cannot be wrong even if they seem to conflict with the evidence.

In Newton's day, his compelling worldview must have seemed completely unassailable. It is an excellent, self-consistent theory which all the available evidence seemed to confirm. But what we often overlook is the extent to which good theories shape their own evidence. By the same token, evidence that conflicts with a cherished theory is very seldom believed. This was demonstrated in Eddington's classic solar eclipse observations to test the theory of general relativity. Although at the time his results were taken to be conclusive confirmation of Einstein's theory, recent analysis of the data indicates that the errors in fact swamp any discernible effects. It seems that the theory's compelling beauty clouded Eddington's judgment and allowed him to conclude that his data confirmed it. This is even a time-honored astronomical tradition. For instance, Ptolemy undoubtedly routinely fudged evidence in order to sustain his geocentric model of the Universe.[30]

49

A more recent example of this tendency is in the influence of DNA forensic evidence, which obtains its validity almost entirely from the commanding power of genetic theory. There is in fact scant evidence for the strongly held belief in the reliability of genetic fingerprinting. Although according to theory there is good reason to maintain that a tiny part of our genetic code is unique to each individual, forensic scientists cannot be sure that they are currently able to identify this individualizing segment since there does not seem to be an adequate data base to support this belief. The element that they claim holds the key to our unique identity could simply be the one that differentiates us into general types. For instance, an expert witness in the O. J. Simpson murder trial claimed that on the basis of a sample of a few hundred African Americans from Detroit she could confidently assert that there is only a 1 in 48 billion chance of another person having the same genetic pattern as that which is in the selected fragment of O. J. Simpson's genetic code. She must have relied on theory and other untested assumptions to validate her seemingly precise probability calculations.

It may be apparent from all of this that the relationship between theory and evidence is complex and confusing and that it is almost impossible to disentangle them or to simply categorize any individual or idea as belonging to one or the other. In consequence I have portrayed Fred Hoyle as both a rationalist and an empiricist, and indicated that what we think of as objective evidence is often an interpretation of the raw data in terms of an existing theory: that theory, so to speak, creates the evidence. We will go into this in more detail in a later chapter, "In the Land of the Blind," which highlights the philosophical differences between Albert Einstein and other quantum physicists. But, for the moment, we will look at the basis of a fundamental disagreement between observational astronomers.

Theory and evidence are, or should be, heavily interdependent. While it may be easy to concoct convincing rationalist theories with very little regard for empirical evidence, the evidence is almost meaningless without a supporting theory. However, although most theorists subscribe to the myth that so-called objective evidence is an independent arbiter of the validity of a theory, experienced observational astronomers are only too well aware of how tenuous and theory dependent their results really are. There is a saying that the only people who believe a theory are the theorists who devised it, and the only ones to disbelieve evidence are those who discovered it. Nevertheless, as we shall see in the story of the tussle between powerful observational astronomers over the value of the Hubble constant, there are important exceptions to this. In the next couple of chapters we will discuss the consequences of the fact that some observational astronomers share the theoretical astronomers' misguided faith in the objectivity and independence of observational evidence.

5

SCALING THE UNIVERSE

For how can you compete,
Being honour bred, with one
Who, were it proved he lies,
Were neither shamed in his own
Nor in his neighbours' eyes?
— W. B. YEATS,
"To a Friend Whose Work Has Come to Nothing"

ONE OF THE GREAT ASTRONOMICAL VENTURES of recent times is the building of the Hubble Space Telescope (HST), a 2.4 meter orbiting telescope designed to obtain images undistorted by the Earth's atmosphere. Ground-based observations are to a greater or lesser extent blurred by air turbulence, which smears fine detail in complex objects, and spreads out the light of faint stars to make them undetectable against the brightness of the night sky. The resolution of a telescope orbiting above the Earth's atmosphere is limited only by the diameter of its main mirror. In the case of the HST this means a resolution of one-twentieth of an arc second, about 10 times better than can normally be achieved from the ground. When the HST was conceived in the early 1980s, it represented a huge advance over any existing optical telescope. However, since the original design was finalized much has happened to ground-based telescopes. For a start, they have become a lot bigger. The largest telescopes in the world at the moment are the twin 10 meter Keck Telescopes in Hawaii. Each of these instruments has a mirror area about 17 times that of the HST, which can go a long way toward making up for the spreading of starlight by the atmosphere. Perhaps more interestingly, it is now becoming possible to correct for the blurring effect of the atmosphere

53

by sophisticated sensor systems that measure the distortions and compensate by bending a mirror in the light path. This technique, known as adaptive optics, is steadily improving the resolution obtainable from the ground, thus eroding the main advantage of the HST. Nevertheless, there are still many areas where the HST is the preferred or the only observational route.

Rather surprisingly, the measurement of the Hubble constant, which sets the scale for the Universe, is not a project for which the HST is absolutely necessary. "Surprisingly," because this was one of the original pillars of the scientific case and indeed almost certainly influenced the naming of the telescope.

Among projects where the HST's resolution is essential is the tracing of the change in the structure of galaxies at higher redshifts. Details of these very distant galaxies are almost completely blurred out by the atmosphere, but the HST has revealed a wealth of structure, some of which is not seen in the local Universe. This is observed in the Hubble Deep Field, a combination of over 300 images of a field only one-twentieth of a degree across, captured over many hours and in several color bands.[1] Clearly, there have been changes in the galaxy population over the last few billion years.

Another exciting new area of discovery which would be essentially infeasible without the HST's resolution is the reconstruction of the mass distribution in clusters of galaxies by observing giant arcs and other distortions caused by gravitational lensing, the bending of light by the gravity of massive compact bodies. From the ground only the crudest features are observable, but the HST reveals a spider's web tracery of fine arcs as the images of distant galaxies are spun out by the massive gravitational effect of nearby galaxy clusters.

The Hubble Space Telescope has produced an impressive library of beautiful pictures, but not all of them have as much scientific potential as the publicity machine would have us believe. For example, the first HST pictures of star-forming regions in giant molecular clouds made spectacular viewing on television news programs, and were accompanied by claims that they had enabled the long-standing problem of star formation to be solved. This of course was quite untrue, but such is the prestige of the HST that it seems every picture must represent a major scientific advance, whereas in this case, it had contributed little more than has been known for the last 10 years. The intimation that this is a major new discovery by HST astronomers is not only dishonest, but is verging on intellectual piracy.

In a similar vein, it seems that nowadays many well-established results require the blessing of the Hubble Space Telescope to be believed. For example, in 1994 I was attending a conference in the United States on low-mass stars, and heard an excellent summary by Jim Liebert of the status of low-mass stars in the galactic halo. He explained in detail the tight limits that had been established over the years on the number of low-mass stars in the halo, underlining the well-known conclusion that no significant amount of mass could reside there in such a form. Imagine my astonishment when, a few months later, a team working with the HST proudly called a press conference to announce that low-mass stars did not make a significant contribution to the mass of the halo. Needless to say, this spurious claim to a new result caused a great deal of annoyance and even anger, but it is by no means an isolated example.

A similar situation arose with the study of white dwarfs in globular clusters, a small contribution by the HST being hailed as a major development. It seems as if no result is now to be believed until it has been attributed to the HST. By the same token any observational claim made by the HST must be correct, or at least more reliable than any Earth-based conclusion.

The most striking example of all this is the HST group's measurement of the Hubble constant. But what is this constant, which shares its name with the world's most prestigious telescope? It is a parameter whose value is perhaps the most important number in astronomy today. When Edwin Hubble realized in the 1920s that the Universe is expanding, he based his conclusion on his observation that, the farther away from the Earth a galaxy lies, the faster it is receding from us. If one makes the additional assumption that the Earth is not in a special place, in other words that Hubble's picture would be seen wherever the Earth happened to be, then it is quite easy to show that the Universe must be expanding. Hubble also showed that the expansion was linear: double the distance of a galaxy, and you double its velocity of recession. The constant of proportionality is what is known as the Hubble constant. Basically, it determines the scale of the Universe. Given a velocity of recession, it tells us the distance of the galaxy.

But not only does the Hubble constant (or H_0, as it is known to astronomers) set the scale of the Universe, it also determines the luminosity of the galaxies within it. The larger the distance scale, the more luminous a galaxy must be for it to appear at the brightness at which we see it. Finally, and perhaps most important of all, in a big-bang Universe H_0 is a measure of the age of the Universe. It tells us the distance to galaxies with a given velocity of recession, so we can work out how long it has taken those galaxies to achieve their present distance from us, assuming that everything was in the same place when the Universe was born. This length of time is then the age of the Universe. The smaller the value of H_0 the larger the distance scale, the more luminous the galaxies, and the older the Universe.

Edwin Hubble tried to find the value of his constant by the same progressive "ladder method" of calculating distance as is used by astronomers working with the HST. This traditional stepping-stone distance scale method involves measurement of nearby stars by triangulation or "parallax," and uses the results to determine the distances to nearby star clusters containing examples of the parallax stars. This then gives us the distance and hence luminosity of all the other stars in the cluster, including luminous variables which can easily be identified at greater distances. Their brightness can then be compared with similar variables within yet more distant clusters, thus estimating the distance of these clusters. Once stars with those same characteristic patterns of variability are identified in nearby galaxies, then the distance ladder is extended still farther. Features of these galaxies are then used to calculate the distance of even farther galaxies, and so on. The more distant the objects, the more steps there are on the ladder, and therefore the more room there is for error.

The first measurement of the Hubble constant in 1927 was 550 kilometers per second per megaparsec. (A megaparsec is about 3 million light-years.) In other words, the velocity of recession of distant galaxies increases by 550 kilometers per second for every 3 million light-years. This gave an age for the Universe which was younger than we now know the Earth to be. But the value was based on the mistaken assumption that a class of variable star seen in nearby galaxies was identical in nature to RR Lyrae variables, a well-studied type of pulsating star. However, between 1945 and 1955 it gradually became obvious to Walter Baade that there was a fundamental error in the original estimate of the Hubble constant and that the variable marker stars were in fact Cepheid variables, which are much more luminous than the supposed RR Lyraes. This had the effect of increasing all extragalactic distances by a factor of 5, and reducing the value of the Hubble constant to 100, close to the current value of 80 found by the HST astronomers.

So, in spite of decades of refinement and the use of the costliest astronomical instrument ever, the value of the Hubble constant has not significantly altered since the discovery of the error in the original measurement. Does this mean that the value of the Hubble constant is becoming ever more secure, steadily converging toward the correct value as the errors are ironed out? Strangely, it does not, as we shall now see, in a saga that flouts all conventional ideas on the structure of scientific debate.

Hubble's great contribution to modern cosmology was to see the importance of finding the value of the velocity/distance ratio, the Hubble constant. He realized that if he could find the relationship between the velocity of recession (which is equivalent to redshift) and the distance of galaxies, then he would be able to directly measure the distance of more remote astronomical objects simply by measuring their redshifts.

The basic requirement for the measurement of the Hubble constant is a knowledge of the redshifts as well as the distances of a selection of objects sufficiently far from the Earth for their motions to be dominated by the expansion of the Universe. (In, for example, the Local Group of galaxies, to which our Galaxy belongs, recessional effects are small and are swamped by the motions of galaxies under their mutual gravitational attraction.) The redshift is the movement of lines in the spectrum of a galaxy toward the red end of the spectrum. This is generally interpreted as a Doppler shift, resulting from the velocity of recession of the galaxy. Most people are familiar with the Doppler shifting of sound waves probably most dramatically experienced when a low-flying ground attack aircraft passes overhead. Providing the aircraft is flying subsonically, which if it is on low-level flying practice it almost certainly will be, a piercing high-pitched scream as the aircraft approaches drops many octaves to a low roar as it flies away. Light suffers the same drop in pitch, or "reddening," when emitted from a receding object such as a galaxy or quasar.

The redshift can be measured quite easily for most galaxies, the main difficulty being to ensure that one uses a representative group of galaxies so that one can average out the effect of random velocities within the group which would otherwise add a "scatter" to the value of the redshift produced by the expansion of the Universe. There are other, more ominous difficulties lurking in the background which could in principle seriously affect the measurement of the Hubble constant. These concern the bulk motions of galaxies on very large scales, which are at present poorly understood. Until systematic effects of this sort have been completely ruled out, it is certainly premature to think of abandoning cosmological models solely on the basis of measurements of the Hubble constant. Even so, the first requirement for the Hubble constant, the measurement of distances to the galaxies, is generally accepted as by far the most difficult part of the procedure.

The earliest assessment of the age of the Universe from the Hubble constant in the 1920s was 1840 million years. Astonishingly, this figure was almost identical to contemporary geological radiometric dating of the Earth at 1800 million years. Whether or not it was pure coincidence that these equally erroneous but otherwise unrelated figures were almost identical, it was understandably seen as brilliant confirmation of the big bang and of the value of the Hubble constant. Two billion years is, after all, a long, long time.

It may be difficult for us to appreciate just how recently we were freed from the biblical method of dating the Universe. Even Darwin belonged to an era when there was heated debate between experts on biblical "begats" as to how many thousands of years ago the Universe came into existence. At that time, physics was brought into the debate only to support lower biblical estimates. Physicists believed that the Sun resembled a great coal fire, in which case there could not possibly be enough fuel to keep it burning for more than a couple of thousand years.

Such theoretical speculations ran counter to the wisdom of the empirical sciences of geology and biology, which required far more time for the evolution of the Earth and its life forms, as they were then understood, to have taken place. Darwinian evolution required millions of years. So even when physicists came up with the cunning idea that, rather than the familiar chemical reactions of an ordinary fire, the energy of the Sun came from gravitational contraction, the mass of the Sun was found to be far too small to provide enough energy through the gravitation mechanism for evolution to have had enough time to take place. The age of the Sun was extended to a few hundred thousand years, which was still hopelessly at odds with the timescales of both geology and biology.

The discovery of radioactivity and the subsequent development of the geological technique of radiometric dating led to a general acceptance by the 1920s that the Earth was billions of years old. But it was not until the late 1930s, a decade after the original measurement of the Hubble constant, that physicists came up with the idea of nuclear fusion as an explanation for the energy output of stars. The Sun was seen as an enormous hydrogen bomb with a life expectancy of 10 billion years. Meanwhile, geologists had established the age of Earth at about 5 billion years, which meant that the Sun must be half way through its life.

Astronomers may have been motivated to find out what was wrong by an awareness of the discrepancy between their estimates for the age of the Universe and the estimates made by geologists and astrophysicists. In any event, they discovered that they had confused RR Lyrae variable stars with the more luminous Cepheid variables. Consequently the Hubble constant was reduced to 100, which implied an age for the Universe of more than 7 billion years. But this did not happen before Hoyle and his colleagues, in order to discredit the big bang, had made much use of the discrepancy between the original Hubble constant measurements and geophysical estimates of the age of the Earth.

The revised value of the Hubble constant implied an age that was consistent with contemporary astrophysics. The fact that the original Hubble constant was based on a simple error somewhat undermined the strength of Hoyle's argument against the big bang. But even this agreement between the findings of distance scale astronomy and astrophysics proved to be ephemeral. In the 1950s, new astrophysical methods for dating globular star clusters in the galactic halo and other astronomical entities implied that the Universe was about 15 billion years old, twice the age that a Hubble constant of 100 would imply. So, except for a short interval, the age of the Universe obtained from measurements of the Hubble constant remained out of kilter with that of astrophysics. Either the Hubble constant was wrong, or there was an increased need for a cosmological constant. The big-bang model was committed to an inelegant ad hoc parameter, and was therefore in trouble again.

However, the strength of the case against the big bang was undermined once more, this time by new procedures for measuring the Hubble constant. Allan Sandage's work in the 1960s revised its value down to between 40 and 50, which was consistent with the finding that the Universe is about 15 billion years without having to invoke the cosmological constant. To this extent the big-bang model was again approaching the elegance required by astronomers such as Hoyle. Nevertheless, it all looked suspiciously convenient that as soon as the value of the Hubble constant proved inconsistent with other observations, threatening the coherence of big-bang theory, it underwent a major revision which brought it into line. In this case the suspicion was deepened by the fact that there was no clear reason to suppose that Sandage's way of calculating the Hubble constant by the distance scale method was any better or worse than that used by those who obtained a value of 100.

Some of the most recent estimates of the Hubble constant still imply an age for the Universe inconsistent with the findings of astrophysics. Observations of globular clusters currently put the age of the Universe at about 13 billion years. Astronomers using data from the HST and Edwin Hubble's distance scale method for measuring the Hubble constant suggest that the Universe is only about 9 billion years old. This is assuming that there is no cosmological constant, and that the critical density of the Universe is in accordance with the theoretical requirements of inflation.

Edwin Hubble was a relentlessly empirical observational astronomer who relied on his own evidence and calculations. He could justifiably claim that his results were relatively untrammeled by the work of physicists or theoreticians, though he appreciated that they needed to be interpreted in the light of theory. The present-day HST astronomers are well within this ruthlessly independent observational tradition perpetuated by influential observers such as Gerard De Vaucouleurs, who, along with his implacable rival, Allan Sandage, was mainly responsible for extending and refining Edwin Hubble's methodology for measuring the Hubble constant.

There is a tendency for observational astronomers to tailor their results to suit widely believed theories. This happens not always because, as Hoyle suggests, astronomers are afraid of finding something new, but because they often lack the confidence to believe interpretations of their own data which conflict with received wisdom or involve a departure from their own established positions. By this I do not mean that they deliberately fudge their data. It is rather that, whenever decisions have to be made about selecting samples, choosing measurement techniques, or discarding apparently flawed data, they will follow their own preferred approach, which is likely to consistently favor a particular outcome. In practice, this generally means that groups tend to confirm their previous results, even though they have new and more extensive observations.

Nevertheless, there is a very strong tradition of independence from theoretical considerations in the community of observational astronomers and such obeisance to the beliefs of theoreticians or the results obtained by other observational methodologies is fairly rare among powerful figures such as De Vaucouleurs or the leaders of the HST group. De Vaucouleurs's attitude was that he did not give a damn what effect his calculations of the Hubble constant had on big-bang cosmology or how much they were at odds with other types of observation. If his results conflicted with the standard big bang, then this model, and not his results, must be wrong. He refused even to discuss the implications of his measurements for the age of the Universe. He did not see theoretical issues as his problem or any of his business. In the same way, theoreticians tend to regard observers as the arbiters of truth and therefore, like a court jury, as inherently infallible. Fred Hoyle is one of the few theoreticians to have seriously challenged the validity of the work of observational astronomers such as Martin Ryle. Although apparently more amiable, the HST astronomers essentially share De Vaucouleurs's independent attitude to theory, and also use the same methodology. However, in my opinion the "ladder" method depends on far too many unsubstantiated assumptions, and the cumulative errors are too great for this way of determining the Hubble constant to be believed to an accuracy much better than 50 percent, no matter how uncontaminated by theoretical considerations it might be.

This is demonstrated by the fact that in the 1970s and 1980s De Vaucouleurs and Sandage each used exactly the same distance scale methodology for calculating the Hubble constant, but their results differed by a factor of 2. In spite of more than 20 years of detailed and ever more sophisticated observations, their respective figures stayed the same, though their estimates of the errors became smaller. This made the differences between their preferred values even more incomprehensible.

It would take a lifetime dedicated to unraveling the complexities of their long-running acrimonious disagreement in order to see who was closer to the truth. However, Sandage had pioneered the independent derivation of our modern value of the age of the Universe from measurements of the ages of stars in globular clusters. In this way he had put the age of the Universe at between 12 and 18 billion years. Consequently, when Sandage came to work on the value of the Hubble constant, he must have been particularly sensitive to the fact that, in big-bang cosmology, high values of the Hubble constant gave an age of the Universe grossly inconsistent with the age range he had calculated. He was strongly motivated to find a consistency with his independent estimate of the age of the Universe. His assessment of a Hubble constant of between 40 and 50 was facilitated by the very ample margins of error of the distance scale method. Because it accorded with the precepts of the big bang and the results of astrophysical age measures, Sandage's value of the Hubble constant was more readily believed than De Vaucouleurs's. Like Martin Ryle, Allan Sandage seemed to win the day over his close rival, but, like the original issues of the big-bang/steady-state controversy, the issue of the value of the Hubble constant has yet to be resolved.

At first sight it is hard to see how two groups using such a similar methodology could possibly come up so persistently with such widely differing answers, and it is interesting to look at how some of their differences arose.

The first step in the distance ladder argument involves the direct measurement of the distance, or parallax, to the nearest stars, with a view to establishing the distance to the Hyades, the nearest star cluster to the Sun. The idea is that once we have the distance to a few stars in the cluster, then we have the distance to several hundred other stars in the same cluster. This sounds straightforward, but immediately we have the sort of problem that bedevils the distance ladder method. The Hyades cluster lies about 160 light-years from the Sun, but is itself some 65 light-years from front to back. (In fact, it has even been speculated that the Sun might be an outlying member of the Hyades.) Therefore, knowing that a star is a member of the Hyades is not sufficient to yield an accurate distance for it. Even the question of cluster membership raises difficulties. The usual approach is to look for stars with a common movement, or "proper motion," across the line of sight, but this is essentially a statistical approach, and is thus uncertain in any particular case. It is generally agreed that uncertainty in the distance to the Hyades translates into a 10 percent uncertainty in the distance scale before we even start, and in my view this is probably an underestimate. Strangely, De Vaucouleurs and Sandage were not in disagreement over this step.

To extend the distance scale beyond the Hyades, we make the assumption that we know the distance of all the Hyades stars and hence their luminosities. We then look for examples of the brightest and most distinctive stars in other more remote clusters. Comparing these stars' apparent and intrinsic brightnesses gives us the distance to those clusters which hopefully contain a large selection of stars, perhaps in unusual and easily recognizable states such as planetary nebulae. These luminous circumstellar shells shed by stars in a late phase of stellar evolution are easily recognizable at very great distances, even in neighboring galaxies. Whether they have a well-defined luminosity is more problematic, but they are nevertheless used as important rungs on the distance ladder.

The next task is to extend the distance scale to nearby galaxies. This is best done by locating distinctive objects such as variable stars, planetary nebulae, and the stars at the high end of the stellar luminosity scale known as supergiants. We can measure the luminosities of these objects from examples in clusters whose distances we already know. This introduces a number of formidable difficulties. Take, for example, Cepheid variables. They are very luminous, and so can be seen in quite distant galaxies, and have a pulsation rate that depends on their luminosity. If one knows this period-luminosity relation, then by measuring the period of variability of a star one can obtain the luminosity and hence its distance. This sounds easy enough, but in practice the relationship is very hard to pin down. The Cepheid variables nearest to the Sun are still a great distance away, so we must rely on intermediate steps to obtain their luminosities. Also, it is now clear that the period–luminosity relation involves other parameters such as metallicity (the fraction of elements other than hydrogen and helium in the star's makeup). A final problem is that the Cepheid variables we can observe in other galaxies tend to be the most luminous long-period examples, whereas the majority of such stars in our Galaxy which are suitable for study are less luminous with shorter periods of variation. So the two groups cannot be directly compared to each other.

In the debate between De Vaucouleurs and Sandage, the choice between the various distance indicators and their relative weighting was a subject of extensive disagreement, and certainly played a major part in the difference between their final results. It also illustrates the limitation of the distance ladder method. Both men made sound cases for their particular approach, but ultimately there was no objective way of deciding between them. It was left to the instincts of onlookers to favor one or the other.

The velocities of nearby galaxies are largely the result of local effects, so to get a good measure of the Hubble constant it is necessary to use much more distant galaxies, whose motion is dominated by the expansion of the Universe. The step from nearby to distant galaxies depends on being able to accurately identify distant counterparts of nearby galaxies whose distance one thinks one knows. There are a number of ways of doing this, all fraught with difficulties.

Much of the Sandage/De Vaucouleurs debate was concerned with identifying features in galaxies which could be used as a measure of luminosity. In practice, this presented endless opportunities for selecting allegedly "unbiased" samples that nonetheless gave each group's preferred result.

More recently, a less subjective approach has been taken which is based on establishing a correlation between the dispersion in the velocities of stars in a galaxy with its overall luminosity. This would enable us to determine the luminosity and hence the distance of a galaxy from measurements of its velocity dispersion. This method certainly has a lot going for it, but still leaves open the question of exactly how the samples are to be selected. Also, it has no real theoretical underpinning. It is essentially an empirical relation which can only be explained in a qualitative way.

It is often said that, even though each step in the distance ladder has an uncertainty of 10 to 20 percent, there is no reason why these errors should all go in the same direction, but they should tend to cancel one another out. Unfortunately even this is not true, as Sandage has been at pains to point out. When we select a sample of distant galaxies to compare with nearby ones, there will always be a tendency to pick brighter ones and miss fainter ones, however carefully sample limits are defined. Consequently we assume that these bright galaxies are representative of an average spread and so believe them to be fainter and therefore nearer than they actually are. This type of selection effect is known as Malmquist bias, and though it can be allowed for, it introduces another insidious uncertainty in the value of the Hubble constant. Sandage's point is that if no allowance is made for the Malmquist bias, the measured value of the Hubble constant will be larger than it should be because many errors will tend not to cancel out but all go in the same direction. I think Sandage is absolutely right to worry about this, even though he has every motivation to do so as it supports his claim for a low value of the Hubble constant. After all, the difference between 50 and 80, the extremes of the currently supported values of the Hubble constant, is only 60 percent, and you don't need many cumulative errors going in the same direction to make up the difference.

Gerard De Vaucouleurs died in 1995, leaving the task of resolving his differences with Alan Sandage to a new generation led by Wendy Freedman, who currently uses data from the Hubble Space Telescope and supports a high value for the Hubble constant. There is a strong feeling in the community of observational astronomers that the HST astronomers have somehow finally managed to succeed where De Vaucouleurs failed. But have they really? What is so special about the Hubble Space Telescope?

Over the course of the Sandage/De Vaucouleurs debate, most of the steps in the distance scale were tied down to an accuracy of 10 or 20 percent. There was, however, one opportunity for a major improvement. Cepheid variables are generally considered to be among the best of distance indicators, in spite of problems with metallicities and so on. However, they could only be detected in the nearest galaxies. The great promise of the HST was to measure Cepheids in the much more distant Virgo Cluster of galaxies, thus cutting out one of the rungs on the distance ladder. In something of an observational tour de force, the

HST has since 1995 yielded some beautiful light curves for these variable stars at incredibly faint levels, and bridged a gap in the distance scale which had been particularly troublesome.

The problem as I see it is that this remarkable observational achievement has clouded scientists' judgment as to its effect on the value of the Hubble constant. Formidable and fundamental difficulties still remain. For example, the Virgo Cluster is a large amorphous aggregation of galaxies which may well be two loose clusters, superimposed along our line of sight, with different galaxy populations. One can even make the case that our Local Group of galaxies, including the Milky Way and the Andromeda Nebula, is an outlying part of the Virgo Cluster. It is very unclear exactly where any given galaxy is relative to the cluster's center, and how the subsets of galaxies which are used to extend the distance scale farther out relate to the subsets of galaxies containing the Cepheid variables.

There are even more disturbing worries. In their 1994 paper in *Nature*,[2] Wendy Freedman's group quotes a value of 80 for the Hubble constant, consistent with earlier high figures published by members of the group. However, if one makes a simple calculation based on the figures given in this paper, one comes up with a value of less than 70. The figure the group quotes is on the upper limit allowed by their errors. I have no doubt that the authors have a satisfactory explanation for this, but it does emphasize the extreme amount of latitude in any calculation of the Hubble constant, even under very restricted conditions.

In contrast to the work of Sandage and his collaborators, these results suggest that either the Universe is considerably younger than the older parts of our Galaxy, or it has a much lower average density than would be compatible with some observations of large-scale structure and, indeed, the theory of inflation. Either several of our methods for estimating lower limits to the age of the Universe are wrong, or the theory of inflation which rescued the big bang from fatal difficulties is a mistake, and the correct picture of the Universe is closer to Einstein's original model. Thus, if we take seriously the value of the Hubble constant suggested by HST results, the average density of the Universe must be too small to be compatible with the theory of inflation. We would need a cosmological constant after all, and with it a renewal of the subculture of theoreticians trying to explain what the cosmological constant could possibly be. This would, essentially, put us back where we were when Hubble introduced his law for determining the distance of remote astronomical objects.

The regressive influence on astronomers in accepting the HST's value of the Hubble constant uncritically was graphically illustrated at a U.K. National Astronomy Meeting in 1996. In a heroic attempt to square the vicious circle of the implications of a high Hubble constant, Martin Rees resorted to the anthropic principle to argue that inflation, which unequivocally requires a critical density and accounts for this state of affairs quite naturally, could nevertheless produce a lower density Universe. Any density other than the critical density is not explained by the inflation mechanism, and can be achieved only if the Universe for some reason decided to stop inflating at exactly the right instant. For it to have stopped at just the right moment to allow the Universe in our epoch to be very nearly but not quite the critical density requires "fine tuning" of a high order, smacking of divine intervention. The anthropic principle to which Rees appealed is a metaphysical argument which takes a number of forms that can best be summed up in the words of Brian Malpass, who explains it as

> the idea that the Universe is the way we see it because we are the ones who are doing the seeing. If that sounds silly, try a large scotch. It still won't sound sensible but you'll feel better about it. There is also a strong anthropic principle, but you don't want to know about that. It's really silly.[3]

As I see it, we are faced with two alternatives. Either we evolved to suit the Universe, or the Universe evolved to suit us. The second option, which is the assumption underlying the anthropic principle, is, to my mind, ludicrous. For a start, it implies that an effect can precede its cause, which is possible only where there is conscious intention or purpose, which in this case requires a divine Creator. (An example of when an effect can be said to precede its cause is my leaving home for Waverly railway station at 9:30 because of the arrival at 10:30 of the train I wish to catch.) Nevertheless, many would argue that intentional behavior is not a proper part of the description of causality, and that "motive" is not the same as either "cause" or "effect." Even so, the anthropic principle reduces to yet another argument from design for the existence of a purposeful divinity.

Rees's argument was basically that when we ponder on the values of the cosmological constants, certain ranges of values are incompatible with a Universe in which there can be observers, and so need not be considered. Quite apart from anything else, this argument relies on a very narrow idea of what would count as an observer, one that would seem to exclude sentient beings of a different order to us such as Fred Hoyle's intelligent molecular cloud from his novel *The Black Cloud*. Even so, the idea can begin to make sense only in a "multiverse" of individual universes, each of which has a random selection of fundamental constants. We may thus ask which of this panoply of universes we inhabit. If, on the other hand, as Stephen Hawking contends, the laws of physics are the ultimate reality and shape the Universe in a way that we can in principle hope to understand, there is no room for the anthropic principle.

In making his case for a low-density Universe, Rees argued that there was no reason for the Universe to have inflated for any longer than was necessary for observers like ourselves to exist. He did not elaborate on the basis of his peculiar "principle of parsimony," which discourages the Universe from inflating more than is absolutely necessary to produce observers. Usually Occam's razor, or the principle of parsimony, is seen as a desirable attribute of scientific theories. This is the first time I have seen it used as a characteristic of the Universe itself, as though it were a natural law. Hopefully there will be no need to develop coherent theories about Rees's fascinating idea that the Universe is ruled by parsimony since some very recent observations suggest that we live in a universe with a critical density and that the theory of inflation as it stands is sufficient to explain why the Universe is the way it is.

All of this is applicable only in the broad context of the big bang. One could take refuge in the steady-state model, where the age of the Universe is not an issue and no Einsteinian cosmological constant is needed to account for the stability of the Universe. So steady-statesmen could claim that the observational evidence gleaned from the HST, though of no particular significance to the underlying rationale for the steady state itself, at least indicates that the big bang still has intractable problems and that the theory of inflation has failed to rescue it after all.

The implications of the evidence from the Hubble Space Telescope are undoubtedly a serious blow to standard big-bang cosmology. This may be a good way of shaking our astronomical mandarins out of their fundamentalist complacency and promoting a greater liberalism in astronomy. However, I believe that in spite of the expensive, glamorous allure of the HST and its beautiful photographs, its astronomers are going down the wrong route to finding the Hubble constant.

THE PRINCIPLE OF
MAXIMUM TRIVIALIZATION

> Here is an example of what seems to be general practice in
> astronomy: when two alternatives are available, choose the more
> trivial. It was so with the discovery of pulsars—white dwarfs,
> everybody said they were, until confrontations with fact showed
> otherwise. And it is so today throughout cosmology. Astrono-
> mers seem to live in terror that someday they will discover some-
> thing important.
>
> —FRED HOYLE, *Home Is Where the Wind Blows*

> Of all mad Creatures, if the Learn'd are right,
> It is the Slaver that kills, and not the bite.
>
> —ALEXANDER POPE, *"Epistle to Dr. Arbuthnot"*

IN THE AFTERMATH OF THE CONTROVERSY between the
steady state and the big bang, it was believed that the main cosmological issues had been solved and the task left to cosmologists
was to clear up the values of a couple of parameters associated with
the big bang. Sandage was at the forefront of those who had settled for
this narrow perception of the remaining cosmological problems. He is
famous for saying that the task of cosmologists reduces to the search
for two numbers: the Hubble constant and the deceleration parameter, which is the geometrical expression of the gravitational force
slowing down the expansion of the Universe. This gravitational force
depends directly on the average density of the Universe, which astronomers express in terms of the density parameter, omega or Ω.

This amounts to saying that all we need to determine is the rate of expansion of the Universe from the moment of creation. If we know the age and average density of the Universe, then we can deduce the linear relation between distance and velocity of recession of distant galaxies, the Hubble constant. Similarly, the Hubble constant and the age will give us the average density or, equivalently, the deceleration parameter, which tells us both how much stuff there is in the Universe and the overall geometry of space. Or we can deduce the age of the Universe from the average density and the Hubble constant.

If one has a preferred value for any one of these three parameters which proves to be inconsistent with what is known to be true of the others, then one can always take refuge in the cosmological constant to correct the apparent inconsistency, provided the value ascribed to it is not too high.

Such circularity and the introduction of inelegant ad hoc parameters like the cosmological constant is characteristic of big-bang cosmology. This can lead to a lot of confusion, especially when provisional models are constructed on the basis of forgotten assumptions, the consequences of which take on the stature of truth simply because they have managed to survive unchallenged for a while.

If we ignore the theory of inflation, and thus the theoretical requirement for the Universe to have exactly the critical density, we must rely on empirical measures of the three cosmological parameters—the Hubble constant, the age of the Universe, and the value of omega—in order to describe the nature of the big-bang universe.

It is exceptionally difficult to measure the average density of the Universe accurately. In practice, we can only obtain lower limits or make assumptions on the basis of theory. However, we have a pretty good idea of the age of the Universe from Sandage's method of determining the age of stars in globular clusters, and from other astrophysical methods such as the cooling time for white dwarfs. We also know that the Universe has to be older than a certain age in order for it, and many of its constituents, to have had time to reach their present state of evolution.

The current methodology for measuring the Hubble constant is, as I have indicated, evidently unreliable despite its apparent attractiveness as the most theoretically independent way of finding any of the three parameters. In De Vaucouleurs's own words (when criticizing Sandage), the distance ladder leaves "much to be desired; it relies on a long chain of assumptions, extrapolations and, in several instances, circular arguments which raise serious doubts about the reliability of their conclusions."[1] Unlike the highly opinionated De Vaucouleurs, at least Sandage and his collaborators had enough modesty to assert that they "have no real way to assess the true systematic error of (their) final value."[2] It seems to me that an observational methodology which produces results at variance with other types of observation, as well as differing from measures using the same methods and which in addition is at odds with theoretical considerations, must be highly suspect whatever the technical excellence of the instrumentation and no matter how prestigious the astronomers involved might be.

This old-fashioned methodology for measuring the Hubble constant not only depends on a number of assumptions, but yields results that are sensitive to arbitrary factors such as decisions about the selection of samples. Because it is a progressive system, sample biases and other inaccuracies can be cumulatively magnified. Although the advantage of having a telescope above the Earth's atmosphere has enabled astronomers to reduce the margin of error of one step in the ladder, they have not advanced beyond the level of sophistication achieved by Sandage and De Vaucouleurs. All that they have really gained is an unsatisfactory compromise between Sandage's estimates of a Hubble constant of about 50 and De Vaucouleurs's result of about 100.

So the old, tedious debate between the virtues of different results from the same unprofitable and unimaginative way of finding the Hubble constant still goes on. No amount of awestruck declamation to spellbound admirers of the Space Telescope can change the fact that the distance ladder method is severely limited. It is no wonder Fred Hoyle thinks that astronomy is moribund and that astronomers are terrified of finding something important.

As a consequence of the findings of the Space Telescope, many astronomers are preoccupied with the regressive business of trying to make the three cosmic parameters balance. Finding compatibility between the Hubble constant, the age of the Universe, and the value of Ω, without recourse to a cosmological constant, is a futile, dead-end preoccupation that characterized big-bang cosmology before Guth brought in his radical solution of the inflation model, thus fixing the value of Ω, albeit only in theory. This theory liberated cosmology from the deadly circularity of early big-bang theory and enabled us to make enormous progress in the 1980s. Now it seems that the Space Telescope is in danger of casting us back into the business of observers bickering over their findings and theoreticians puzzling over the fundamental tenability of the big-bang idea.

Although it is more sensible to suspect the validity of the value of the Hubble constant which has emerged from the HST, some astronomers prefer to regard it as inviolate and to question the more secure observational evidence for the age of the Universe, and also the less secure evidence that the Universe has the critical density. Although insecure because our way of finding the total amount of matter in the Universe, and hence how much is unaccounted for, is also fraught with uncertainties, the current methodology for finding the value of Ω could arguably be seen as no less reliable than the distance scale method for finding the value of the Hubble constant.

The most frequently used approach to the measurement of the density parameter Ω typically involves analyzing the way galaxies are clustered. The idea is that the degree of clustering we observe is a measure of the amount of mass in the Universe. As the Universe evolves, a high mass density will result in large concentrations of material. In principle this is an excellent method and should be capable of giving an accurate value for Ω. Unfortunately, one critical assumption must be made. When we measure the clustering of galaxies, are we measuring the clustering of mass, or could the underlying dark matter have a different clustering, perhaps on a larger scale? There is no way of answering this question at the moment, though there are a number of ideas floating around as to how this "bias" might be measured.

A new and potentially very powerful approach involves measuring the fluctuations in the microwave background radiation on smaller scales than was achieved by NASA's Cosmic Background Explorer satellite, COBE.[3] The first results from this endeavor were obtained by radio astronomers at Cambridge using a ground-based array of microwave detectors.[4] One can predict from theoretical considerations that the fluctuations over different angles on the sky will be different, and that the exact shape of the "spectrum" of fluctuations will depend strongly on Ω. It is actually not too different from the idea of looking at clustering of galaxies on different scales. The new measurements obtained by the Cambridge group show a particular feature in the spectrum just where expected for a critical-density Universe, but at the moment their errors are too large to be sure that a low-density Universe is excluded.

Another method for measuring omega goes back to an approach used by Allan Sandage and others in the 1960s. The idea is to make a more direct measure of the geometry of the Universe which can then be related quite simply to Ω. If the Universe has the critical density, then its geometry will be flat. If the value of Ω is either larger or smaller than the critical density, then the overall shape of the Universe will be curved into a shape analogous to either a sphere or a saddle. All this is another way of saying that the Universe is either flat, closed, or open.

Suppose we wished to know whether we lived on an infinitely flat Earth or on a sphere; we could use a straightforward geometrical test. It is possible to map out a huge triangle on the Earth's surface by defining three points and constructing the sides as the shortest routes running along the surface between the three points. If the Earth were flat, the sum of the angles of this triangle would exactly equal 180 degrees. If, instead, we constructed the triangle on the surface of our Earth, we would find that the sum of the angles came to more than 180 degrees, and with more detailed measurements we could work out the Earth's exact shape. Sandage set out to use this method to find the geometry of the Universe, but his program eventually foundered on the difficulty of knowing whether or not the very luminous galaxies and quasars he was using as markers for his gigantic triangles had luminosities that were constant throughout space and time.

Recently this method has been given a new lease on life. In place of galaxies, it has become possible to use supernovae as markers. The technique is to discover supernovae at great distances in time to monitor their changes in brightness so that they can be classified and their luminosities estimated. There is a strong case that the intrinsic brightness of supernovae will not change as we look back in time. This, along with their enormous luminosities, makes supernovae excellent markers for mapping the geometry of the Universe. It is still a little early to assess the results of this new test, but already it seems that a Universe with a cosmological constant is inconsistent with the data.

By accepting a Hubble constant of about 80, astronomers have painted themselves into the corner of having to accept that the theory of inflation, and therefore the big bang, is falsified. In consequence they are resorting to desperate measures symptomatic of the death throes of a dominant theory. They are trying to modify inflation so that it no longer requires the Universe to have the critical density. Astonishingly, even though the critical density is fundamental to inflation, such tinkering is taken seriously. This is regardless of the fact that the very problem that inflation solved, impossibly balanced ad hoc initial conditions, has to be invoked to explain the revised model of inflation. To my mind, this illustrates the danger of blind reliance on observational evidence to test a theory. History has shown that observations are just as likely to be wrong as theories. Also they are undoubtedly harder to challenge because, as the De Vaucouleurs/ Sandage squabble illustrates, it is very difficult to argue with the technical minutiae that go into producing observational evidence, especially when orchestrated by astronomical leaders of the day.

Also, it takes almost as much effort and courage to question the competence of a time-honored observational methodology as it does to overturn an orthodox cosmological model. Everyone not intimately involved in such an affair is reduced to the status of helpless onlooker. Similarly, most of us feel powerless to question the confident findings of those using the HST to determine the value of the Hubble constant. This is regarded as the hard evidence which can be used to overthrow the best theories and which every theoretician dreads.

The allure of the Hubble Space Telescope has engendered in those who work with it an almost unquestioning acceptance of their results. As is true throughout astronomy, most of the observers working on the HST data will be postdoctoral researchers who, if they want to remain in astronomy, would be reluctant to interpret data in any way that deviates from the views of their superiors. Consequently the leadership's confidence in their methods will increase, and so a vicious circle is set up. Meanwhile, those of us looking on can marvel at the remarkable agreement between their results, and those who are less cynical than I am will become convinced that the HST value of the Hubble constant must be right.

In my opinion, any observation that conflicts with the theory of inflation also conflicts with big-bang cosmology, and should either be regarded in that light or disbelieved, no matter how much power those making the observations might have. Actually, powerful groups are especially suspect. The more powerful the leaders are, the more difficult it is for the students and postdoctoral researchers to depart from the party line. Survivors in the field of astronomy not only avoid any work that might be seen as dissent, but cannot afford even to mention dissident ideas. Those who are really wise won't even think such thoughts. I do not think much can be done about this, but in the long run I am not sure that it really matters. As long as everyone understands the ground rules, and competing groups are pushing different lines, the untenable ideas will eventually become apparent. A problem will arise only when one group monopolizes the field and the possibility of bias is not admitted.

I have so far painted a rather gloomy picture of the situation regarding the measurement of the Hubble constant. Using the distance ladder method we are not converging toward a single value but still oscillating between values that differ by a factor of as much as 2. The only change seems to be in the error estimates. It seems clear to me that the distance ladder method reached the limit of its usefulness a long time ago. In fact, it is remarkable that it has produced agreement to within a factor of 2. There are not many cosmological parameters known with such accuracy.

In my view it will only be with the application of new, single-step methods that we shall ever learn the value of the Hubble constant to within 10 percent. As with the attempt to find the value of Ω by geometric methods, the most straightforward of such single-step methods for finding the Hubble constant involves supernovae explosions, when for a few days a single star can outshine a galaxy. Such extremely luminous events can be observed at huge distances, easily far enough for them to be reliable measures of the Hubble flow, as the smoothed-out recession of the galaxies is known. The difficulty is to know the intrinsic luminosity of supernovae sufficiently accurately. The problem is particularly hard since they come in a variety of types, each with its own characteristic luminosity. So far this problem has not been solved, but when it is, it will enable us to leapfrog most of the steps in the distance ladder. Interestingly, the best results achieved with the supernova method so far suggest a value for the Hubble constant of around 50, much lower than the HST value.

A completely different approach to measuring the Hubble constant involves the phenomenon of gravitational lensing, when light is bent by the gravity of a massive object. We shall explore this in detail later in this book. The manifestation of gravitational lensing in which we are interested here involves the very compact luminous objects known as quasars.

When these objects were first discovered by Maarten Schmidt in 1963,[5] the extreme nature of their observed properties posed a real conundrum for theoreticians to disentangle, and even today there are many aspects of quasars which are not understood.

The most striking feature of quasars is their large redshifts, which if interpreted as part of the Hubble flow, imply that they are very distant, billions of light-years, and hence to achieve their observed brightness they must be extremely luminous. Furthermore, quasar images are essentially starlike, implying that they are very compact. They appear as clearly defined pinpricks of light rather than the more nebulous smudgy images of less compact entities such as galaxies. Such compactness suggests that they are fairly small. This is rein-forced by the observation that they can vary in brightness over a few months, which means that they cannot be more than a few light-months across, not much bigger than our Solar System.

For a quasar to change brightness by a significant amount, there must be coherence from one side to the other. Suppose some distur-bance on the quasar disk triggers the variation. The outer parts can only respond after a finite amount of time, and this can never be less than the time it takes light to traverse the distance from the distur-bance. For a quasar a few light months across, this will be a few months. Thus if we see a quasar varying on a timescale of months, we can be sure that it is no more than light-months across.

The most baffling part of the quasar conundrum was how such a vast amount of energy, equivalent to the light from a hundred gal-axies, could be generated in a volume the size of the Solar System. Among early attempts to explain quasars was the denial by Martin Ryle and others that they were at the distances indicated by their redshift. If they were much closer, they would be much less luminous. This raised the question of what caused the redshifts, and led to the idea of "discrepant" redshifts. Halton Arp pointed to a number of apparent associations between high-redshift quasars and low-redshift galaxies, claiming that the objects were connected by streams of matter and that the redshifts must have some explanation other than the expansion of the Universe.

In the 1970s there was extensive debate over the question of discrepant redshifts between Arp and John Bahcall, who championed the cosmological interpretation. This debate was characterized by Arp's conviction that discrepant redshifts provided a clear refutation of an expanding Universe and the associated big-bang picture. Bahcall and the defenders of orthodoxy concentrated on picking holes in Arp's statistics, which was on the whole not too difficult. Although the orthodox view carried the day, and Arp now claims that it is impossible for him to even publish a paper supporting discrepant redshifts, a number of nagging loose ends remain.

In my view, Arp adopted a somewhat naive strategy in the debate. He was clearly affronted by the fact that what he considered to be solid evidence in favor of discrepant redshifts, which would certainly have been regarded as sufficient to make a more conventional point, was either dismissed on the basis of minor statistical quibbles or completely ignored. What he did not seem to appreciate was that, without a theoretical framework capable of accounting for all generally accepted observations and making verifiable predictions, critics of his ideas were free to attack his observations case by case and write off any residual effects as statistical flukes.

There has in fact been some attempt over the years to devise a theoretical backing for discrepant redshifts. Fred Hoyle explored the possibility that they might be caused by extreme gravitational fields, and Arp himself investigated the idea that redshift decreases with age,[6] but I think it fair to say that no alternative theory has been proposed which compares in elegance and comprehensiveness with the orthodox view of an expanding Universe.

Ironically, Hoyle's interest in a gravitational explanation for discrepant redshifts was preceded by a paper, published shortly before the discovery of quasars, in which he postulated the existence of gravitationally collapsed objects producing huge quantities of energy.[7] Hoyle has since claimed that this paper constituted a prediction of the existence of quasars. Although this might be an overgenerous interpretation of what is actually stated in the paper, it is certainly true that Hoyle's point about extracting large amounts of energy from gravitationally collapsed bodies has formed the basis of our understanding of how quasars work.

Our current picture of quasars is actually quite complicated. In the center of a galaxy lies a massive black hole containing around 100 million Sun's worth of matter. This black hole forms at the center of a vortex or whirlpool of matter spiraling inward. Surrounding matter is drawn into this accretion disk where it is accelerated to extremely high speeds before disappearing into the black hole and losing all contact with our Universe. During the final moments before crossing the event horizon, when traveling very close to the speed of light, the matter can emit as much as half of its mass in the form of electromagnetic and other radiation, the whole way across the spectrum from microwaves, through the optical and ultraviolet, to X-rays and gamma rays. Enveloping the accretion disk are clouds of gas which are stimulated to shine, rather like fluorescent lights, by the intense energy pouring out from the center. The picture is completed by two opposing jets of electrically conducting gas or "plasma" ejected from the center, perpendicular to the disk. The entire accretion disk system is smaller than our Solar System, and so can vary on timescales of a few months, as required by the observations. Quasars are discussed extensively in a later chapter; for the moment it is interesting to see how these fascinating objects form the basis for one of the most promising methods of measuring the Hubble constant.

When a massive galaxy lies along our line of sight to a compact light source such as a quasar, the gravitational field of the galaxy bends the quasar's light rather in the manner of a glass lens. The effect on the quasar is typically to split its image so that one sees two quasars separated on the sky by a few seconds of arc, and differing only in their apparent brightness. Now, if the quasar varies in luminosity, for example if it gives off a flare, this will be seen in both images but not necessarily at the same time. The two sets of light rays from the quasar follow different paths through the Universe to the observer, and typically these paths will not be the same length. Thus one might observe the flare in one image several months before it is seen in the other. Given that we know the speed of light, the time difference between the two detections gives us the difference in the path lengths, which in turn gives a direct measure of the Hubble constant. The only other requirement is that the mass distribution and redshift of the lensing galaxy must be known, since these can affect the length of the light paths. So far only one double quasar system has been discovered which is suitable for measuring the Hubble constant, and even here there is some uncertainty about the mass distribution and the time delay. Still, I have no doubt that as more systems are discovered and monitored for variations a solid value for the Hubble constant will emerge. Thus it seems to me that quasars are at present the most promising route to an independent measure of the scale of the Universe.

From a purely academic point of view, finding the Hubble constant by direct methods such as the distance scale or double quasars is unnecessary. In principle its value can be deduced from the age of the Universe and the value of Ω. Observational estimates for the age, though still open to debate, are arguably more secure than the distance scale measures of the Hubble constant and there is evidence from large-scale motions of galaxies and galaxy clusters that Ω is close to the critical density. The theory of inflation, which unequivocally requires the critical density, lends theoretical weight to the larger observational estimates of Ω.

It is nevertheless a deeply ingrained instinct for scientists to seek incontrovertible evidence for or against an idea, no matter how firmly it is believed, and no matter how reliable, say, computer simulations might be. There is no substitute for observational evidence or controlled testing. I remember asking my French collaborator, Philippe Véron, why on Earth France was prepared to stir up such a hornet's nest over its determination to carry out nuclear tests, when surely there could be no real doubt that the bombs worked. He replied that we knew our quasar candidates all had extremely high redshifts, but we still wanted to measure them directly, just in case!

My concern is that we should adopt more productive and imaginative methods for finding the Hubble constant than the outdated and irreducibly unreliable distance scale method, which I feel has reached the limit of its accuracy as a measurement strategy. It is far too important to be left to a powerful minority of traditionalists. We should at least make use of new ideas and discoveries. There has been no progress in refining the Hubble constant, even though we have been at it for three quarters of a century. Viewed in this light, it seems that Fred Hoyle does have a point when he claims that, apart from the theory of inflation, big-bang cosmology has not made any significant advances or important discoveries since the 1960s.

7

IN THE LAND OF
THE BLIND

God guard me from those thoughts men think
In the mind alone
> —W. B. YEATS, *"A Prayer for Old Age"*

BECAUSE I BELIEVE that an understanding of some of the fundamental debates of philosophy is crucial to an appreciation of the scientific process, I think it would be a good idea at this point to look in greater depth at what underlies our idea of reality. So, before going into all the specific ideas that make up our current model of the Universe, it is worth trying to get a deeper feeling for the issues involved in the conflict between rationalism and empiricism and hence the relationship between theory and evidence, which together shape our picture of the objective world. I think we can get a fairly coherent view of what this is all about by discussing an aspect of this conflict which has created a deep philosophical division between cosmology and quantum physics. This dates from the time of Einstein's outraged repudiation of the conceptual basis of quantum mechanics, and his decades long argument with the nuclear physicist Niels Bohr.[1]

At first, Einstein was impressed by the apparent rigor and potential certainty of the form of empiricism suggested by the Vienna School of philosophers, the logical positivists,[2] which numbered among its disciples the father of quantum physics, Werner Heisenberg. Nevertheless, Einstein came to appreciate the serious limitations of logical positivism when Heisenberg put forward his uncertainty principle, which Einstein felt violated the most fundamental tenets of reasoning.

Put simply, logical positivism is a type of extreme empiricism which prescribes the construction of scientific ideas out of accumulated observational data. Theories can only be regarded as scientific if one knows how to go about finding tangible supporting evidence. Any proposition, such as the existence of God, which cannot be verified by experience is quite literally nonsense, according to the logical positivists.

However, Einstein felt that the empirical rigor of positivism simply leads to worse nonsense, with the added problem of being irrational and incoherent. Although the story of Schrödinger's famous indeterministic cat[3] was intended as a lighthearted *reductio ad absurdum* of the ontological implications of quantum physics, it nevertheless characterizes the sort of craziness that Einstein felt they were committed to.

Heisenberg's uncertainty principle formed the conceptual basis of quantum mechanics and led to the indeterminist epistemology formulated by physicists such as Niels Bohr and Erwin Schrödinger, which Einstein thought of as little more than sophistry. Nevertheless, the influence of logical positivism on modern science and on scientific philosophizing is far from negligible.

The philosophical split between Einstein and Heisenberg characterizes the main polarities of current controversy over the relationship between theory and evidence. Einstein felt that rationalist theories were far more likely to lead to progress than any inductivist assemblage of experimental data. He maintained that the radical skepticism which is intrinsic to empiricism inhibits scientific progress. If one is skeptical even about the notion of objective reality itself, then since we cannot prove that what we experience bears any relationship to what may be external to our minds, what is the point of scientific investigation? The logical conclusion of such doubt is the denial that we can ever know anything since even our most direct experiences may be nothing more than a set of subjective illusions. Thus, far from leading to the scientific certainties which positivism claimed to deliver, it leads to far greater uncertainty than any rationalist epistemology, which at least maintains some primitive assumptions that do not need proving. Such unchallengeable premises include the notion that there is independent external reality that we can access and understand. This is an intellectual certainty that does not need empirical proof.

As Wittgenstein himself realized, there are some ideas that it does not even make sense to challenge. There are some questions that are unanswerable.[4] For instance, just as an eye cannot look directly into itself, so we cannot by means of our built-in idea of objective reality decide whether it is in fact objective. By what criteria could we determine whether or not our belief in objective reality is justified? Even if there were such criteria, how would we know that they are themselves valid? Such empirical questions are as meaningless as those posed by rationalism. This is not to say that they are self-evidently true or false, but simply that these terms are not applicable. Such ideas are not susceptible to analysis. We should not ask rationalist questions like, "Does God exist?", or empirical questions such as, "Is there external reality?" since we would just end up talking nonsense. For this reason, Wittgenstein introduced the dictum that analysis has to stop somewhere: "What we cannot speak about we must pass over in silence."[5] If something is imponderable, we should not dwell on it unless we wish to talk metaphysical guff. We should accept that there are fundamental limitations on what we can know or understand, and that part of the task of philosophy is to recognize just where these conceptual boundaries lie.

In order to do this philosophers cannot examine objective reality itself. This would be tautological, or at least vacuous, since it would presuppose what it is that they are trying to ascertain. The only fruitful way of investigating the objective world is by looking at our relationship to this reality. We must look to our existing conceptual frameworks or our collective mind-sets. Since we are not mind readers, the only way to access our collective mentality is via our common language.

But Wittgenstein goes further and maintains that our ordinary language not merely represents, but *is* objective reality since the attempt to distinguish our ideas from our use of language is as meaningless as trying to distinguish our ideas of the objective world from the objective world itself. Again, by what criteria could we decide that our idea of reality corresponds to the meaning and structure of our language? It is like asking how I can know that my idea of red corresponds to the word "red." This is one of the fundamentally intractable questions that it does not make sense to ask, unless one interprets it simply as an inquiry into whether or not my use of the word "red" corresponds to the way most people use it. Wittgenstein felt that it is meaningless even to make the statement that our use of language is indistinguishable from our idea of reality, which is in turn indistinguishable from objective reality itself. This is because it is addressing something that "we cannot speak about." Accordingly, he says that "anyone who understands me eventually recognized them [his propositions] as nonsense."[6]

Niels Bohr shared Wittgenstein's view that there are fundamental limitations to the scope of human knowledge. He wrote, "We must not forget that, in spite of their limitations, we can by no means dispense with those forms of perception which color our whole language and in terms of which all experience must ultimately be expressed."[7]

Modern astronomy describes the Universe in terms of the Platonic realism of Newtonian physics and Einstein's theory of general relativity, while quantum physics views it in terms of the extreme empiricism of Heisenberg's uncertainty principle. Although these radically different worldviews have not so far been reconciled and are possibly irreconcilable, they have proved outstandingly successful within their respective domains. On the whole, the ideas and theories of these very different scientific cultures do not rival each other. They are not mutually exclusive, and therefore cannot supplant or refute each other, but have an uneasy coexistence. It is only where there is some overlap that paradoxes and incoherence occur.

Because of a refusal to accept these inherent differences and the resulting attempt to reconcile the ideas of quantum physics with classical physics, a number of counterintuitive and philosophically untenable ideas have achieved some prominence. This is the case with some theories that try to straddle astrophysics and quantum physics, such as the attempt to describe the whole Universe in terms of the behavior of elementary particles by concocting the cosmology of parallel Universes. This cosmology admits every possibility into reality. It solves the problem of nothing existing except in our minds, simply by allowing everything to be objectively true. It is just that some events, or universes, are more true than others, and are therefore more likely to be actualized. But this solves absolutely nothing. Apart

86

from substituting the term "actual" for "real," there is in essence no difference between the two views. To say that every possibility is real is as vacuous as saying that nothing is real. In order to save either epistemology from complete vacuity or meaninglessness, we have to rely on our subjective experiences either to decide what can be admitted into reality or, in the case of the many-worlds cosmology, the degrees of this reality and hence the probability of anything being actualized in our Universe.

I believe that the attempt to reconcile the Platonic realism of classical physics with the inductive empiricism of quantum physics will always be philosophically untenable until theoreticians can talk about and agree on their fundamental philosophical positions, and at least recognize that they are playing different games. As it is, there are a few lost souls occupying this area who do not know what game they are playing. They can be heard pontificating fervently on the esoteric and perplexing nature of reality while wandering in the intellectual labyrinths of no man's land between astrophysics and quantum physics.[8]

It seems to me, despite attempts at realist interpretations of the observations of quantum physics with the consequent confusion, that nevertheless quantum physics is still inherently Wittgensteinian, and sometimes makes more sense than classical physics. There is an influential element within quantum physics which maintains that there are fundamental boundaries to knowledge, and that therefore there are some questions being asked which are either vacuous or meaningless.[9] Also, there are some interpretations of quantum data which seem to contradict the apparently inviolate laws of classical physics. For example, some data suggest that the second law of thermodynamics is far from inviolate, and that order can be obtained from chaos.[10] This certainly accords with our everyday experience and, if true, we would no longer have to categorize life and the activities of life forms as pathological exceptions to a law of nature, nor would we get into quandaries (brief or otherwise) about the nature of time. Thus quantum physics has the potential to finally liberate us from the linearity of Newtonian mechanistic determinism which still dominates our way of thinking about the Universe.[11] Newtonian physics pictures the Universe as a sort of clock that is wound up at the beginning and then gradually runs down and wears out. This concept, along with the urge to date the Universe, is an extension of the biblical idea that the

Universe had a beginning and will have an end, and that it was consciously designed. The deeply entrenched biblical notion of the linearity of time carries with it the idea of increasing entropy or chaos, which, in turn, presupposes that the Universe is analogous to an artifact and that therefore it must have been created.

If one can dispense with the second law of thermodynamics and like our pre–Judeo-Christian ancestors regard time as cyclical, see everything as being in constant flux between chaos and order and back again, and apprehend nature as constantly creating and recreating itself, one opens the door to a completely different physics. The steady-state cosmology proposed by Fred Hoyle and his collaborators is one of the best modern examples of such a maverick challenge to the Newtonian biblical idea of the linearity of time and the validity of the law of entropy. Nevertheless, since it is more a rationalist than an empirical idea, it was relatively easy to defeat by the equally rationalist arguments for a big-bang Universe which embraces the biblical notion of a beginning to all creation, and has embedded within it the perceived inviolacy of the law that everything is degenerating into chaos and eventual extinction.

The only way for the challenge to Newtonian physics to become anything more than a nice thought experiment is by strictly empirical science, such as early quantum physics or Darwinism. Such science is relatively unencumbered by rationalist ideas deeply rooted in our culture, and therefore by the self-defeating compulsion to try to embed its results within the framework of classical physics or in Darwin's case, biblical history. (Darwin weakened his argument considerably by his well-mannered attempt to reconcile his principle of natural selection with Christian dogma. However, his less scrupulous colleague, T. H. Huxley, eventually made Darwin face up to and embrace the frightening and liberating fact that his ideas were irreconcilable with the world order of the Victorian age.)

So, what is the exact nature of this great philosophical disagreement that has had such far-reaching consequences for physics? At the risk of oversimplifying, I think it can be summed up as a conflict between natural inductive, or commonsense, reasoning and the artificial deductive reasoning of formal logic and mathematics.

Natural reasoning, which is reflected in natural languages, represents the complicated way we ordinarily go about making sense of our environment. Scientific thinking is in part a more systematic and disciplined version of such inductive commonsense reasoning, particularly when gathering evidence in support of scientific models. However, the scientific method is largely a system of self-sufficient ideas that can be coherently expressed in the artificial languages of deductive logic or mathematics.

Theoreticians regularly construct entire cosmologies with very little reference to experimental or observational evidence. This is especially true of the ancient art of astronomy, which started to take observational evidence seriously only as recently as the Renaissance. Even so, the old ontological tradition of rationalism still persists among astronomers. Rationalism is the belief that the higher intellect is a more reliable path to apprehending reality than commonsense thinking or what we can experience. Thus there is the feeling that what is intuitively obvious or empirically derived often bears very little relationship to what is really the case: that intellectual or mathematical models are more likely to provide us with an idea of the essential nature of reality.

This conflict between empiricism and rationalism is part of a general epistemological problem. How can we have confidence that our theories and perceptions of the way things are bear any relationship to what is really the case in the objective world? In what way do scientific theories represent reality? Is it valid to interpret our models as true descriptions of nature, or should we see them as no more than metaphors for abstract intellectual ideas?

Particle physics is a very young discipline, and consequently relatively unencumbered with traditional ways of interpreting reality. Thus quantum physics took the modern empirical view that it is invalid to ascribe qualities to the Universe which we cannot observe. By the same token, we have no alternative but to assume that the qualities we do observe really exist. What we see and the way we see it is what there is. What cannot in principle be observed has no existence. Another way of expressing this idea is to say that a statement cannot be regarded as being about matters of fact and existence unless we know how to go about verifying this statement. As long as there is no possibility of verifying or falsifying an hypothesis, it will not be scientific. In other words, unless you can test the validity of a statement by experiment or observation, it has no scientific meaning.

It is in principle impossible to accurately measure both the velocity and the position of an elementary particle such as an electron. Some form of electromagnetic radiation, such as visible light or X-rays, is necessary for the measurement. All electromagnetic radiation comes in discrete, indivisible parcels called quanta, which have various wavelengths. These quanta will always interfere with either the velocity or the position of the electron, depending on their wavelength. The shorter the wavelength, the more accurately quanta can be used to measure the position of the electron. However, shorter wavelengths correspond to higher energy levels, which interfere more with the velocity than those with longer wavelengths, which are, conversely, less accurate measures of position. It is analogous to shining light on something in order to see how dark it is. More generally, light always exerts some pressure on physical objects. At subatomic levels this pressure is very significant. The only way to completely prevent light from altering the behavior of such tiny objects is to turn off the light. But then you can't observe anything. There is absolutely no way out of this impasse. It is fundamental. In other words, nature is such that it will always be impossible to accurately observe both the velocity and the position of an elementary particle.

We are thus limited in our ability to make predictions about the behavior of subatomic particles. We can only make probabilistic predictions about such entities. Consequently we think of them as having a free, indeterminate sort of existence which is rich in potential but lacking in definition. We have to regard them as having their being in this mixture of possible positions and velocities, or "quantum states," because it is the only information about them that we can ever get. They take on definable attributes and therefore a fleeting identity only at the instant of being observed. It is as though only the act of observing them realizes their potential and brings them into existence, rather in the way that the redness of an object comes into being only when someone who is not color blind sees it. Thus the observer's intimate participation in the nature and reality of the objective world is crucial. It also shows that there is a fundamental limit to what we can know. There could be any number of hidden variables that might explain the apparent indeterminacy of subatomic particles, but we can never know them, which is the same as saying that they do not exist. If they do not exist for us, then they do not exist at all. We simply cannot coherently discriminate between the thing as it is in itself and the thing we can apprehend. The limit of what we can know is also the limit of reality. To postulate hidden variables or other "super-phenomena" is a rationalist fallacy that not only often leads to contradiction, but also serves to trap us within a conceptual framework.

Alternatively, one could adopt the extreme realist position held by some present-day physicists who straddle both astrophysics and particle physics, and try to devise models of the Universe based on observations of subatomic particles. The only way to give these phenomena an observer-independent reality is by thinking of them as existing simultaneously in all their possible states. Thus all their possible paths and positions have a timeless existence. It is just a commonsense notion to discriminate between events that might happen and those that have happened.

As we have seen, this rationalist model based on the apparent ability of elementary particles to occupy all positions at the same time in our Universe can be expressed in terms of parallel universes, or a multiverse. All possible universes are real. All possible states of affairs not only will be, but are and have been. Tense is an illusion, and even words like "now" have no meaning in this model of reality. These otherwise discrete universes intrude on our Universe at the subatomic level. So presumably all possible universes must intrude on each other all the time. But it simply does not make sense to postulate an infinity of parallel universes on the basis that all universes are somehow present in one Universe at the subatomic level. This idea is a mathematically coherent way of providing a solution to the problem that arose from the apparent ability of subatomic particles to occupy an infinity of positions at the same time. But, as J. S. Bell says, "If such a theory were taken seriously it would hardly be possible to take anything else seriously."[12] To my mind, this many-worlds idea is a classic manifestation of the sort of nonsense that Wittgenstein warns us against.

The many-worlds hypothesis is a good example of the paradoxes and consequent mystification that often result from trying to interpret mathematical models in terms of reality. It may be valid to express our ideas of reality in mathematical terms, but does it always make sense to do the converse? Analogously, we can sensibly discuss the extent to which reality is reflected in a representational painting. On the other hand it often does not make sense to try to interpret a purely abstract painting as reality. If it is not depicting anything beyond itself, it is pretentious to try to imbue it with meaning. However beautiful it may be, it is not describing reality except to the extent that it is itself part of the world. It means little more than say a lovely color. Thus abstract art becomes incomprehensible and elitist in the hands of critics who perpetuate the fallacy that because something seems to mean very little, it must mean a great deal; that because abstract art represents nothing tangible, its elegant simplicity must represent the deeper and higher truths of a sort of emotional and intellectual hyperspace which are too complex for mere thought or words, and are certainly beyond the common experience of untrained ordinary people.

Another way of expressing the rationale for postulating a multiverse is to say that since, in Einstein's words, "God does not play dice," he must allow all possibilities a real existence. Also, since God created everything, he must be independent of his creation and therefore apprehends the Universe as containing all possibilities simultaneously, without the constraints of the linear space-time continuum that we mere mortals are trapped within, and which we can only have the illusion of transcending within our inadequate imagination or defective memories. Formal logic, especially mathematical physics, is the only way of thinking that might give us some clue to the essential nature of reality, to the way things really are. We cannot get beyond the limitations of our senses and psychology with commonsense or inductive reasoning. This merely gives us illusions such as the feeling that time has a direction and obscures our apprehension of the timeless reality provided by mathematical reasoning. We are simply wrong when we make a statement such as, "It is possible for ravens to be red, but there are no known instances of red ravens, therefore red ravens do not exist." In other words, if a statement has meaning then it is describing reality regardless of whether or not one can find any verifying instances. Reality transcends contingency. However, this holistic argument is fundamentally incoherent. How could we experience the direction of time if it is not real? How could we experience anything? Our ideas of causality and identity which we need to make any sense of the world would become meaningless. All our ideas of reality would be based on fundamental illusions. Scientific endeavor would become pointless.

However, very few of us adopt either of the extremes of rationalism or empiricism. For instance, most of us think of colors as properties of the real world, specific to time and space, which are part of the description of unique individual objects having a finite, contingent existence regardless of whether or not they are observed. We say, "That object is red," not, "That object has properties which affect my eyes in such a way as to make my brain come up with the idea of redness," nor, "That object appears red to me, but it could be any other color, therefore it is really all colors and no color."

Empiricism is not confined to the existential status of colors or elementary particles. According to this philosophy, everything is a type of illusion. We cannot describe real properties but only the effect the world has on our senses and therefore on our psychology. We cannot in principle get beyond the information provided by our five senses, so we can never know how the external world is in itself. In effect, the objective world does not exist except as a potentiality which we bring into being when we observe it. But because we can never get beyond the constraints of our senses and mind-sets, we simply have to accept that what we observe is real. This, to us, is the real world, and it is pointless to speculate about what is inaccessible. To us, the Platonic Universe is not only useless and unknowable but completely meaningless.

Rationalism also takes the view that we cannot access absolute reality through sensory experience. We can, however, obtain some idea of the essential nature of the Universe as it is in itself by thought processes which we encapsulate in logically transparent languages such as formal logic or mathematics. It is only in this way that we can, to use a degraded metaphor, know the mind of God. Sensory information and the related commonsense inductive reasoning leads us into error and delusion, which is the currency of dreams and cannot claim the status of representing reality.

A useful analogy for the conflict between the ontologies of rationalism and empiricism can be drawn from the well-known dictum that in the land of the blind the one-eyed man is king. In H. G. Wells's story *The Land of the Blind,* the sighted intruder into a totally isolated blind community is treated as mentally ill by their scientists and civic leaders, while a few of the more imaginative and mystically inclined people regard him as something of a visionary. Relative to the blind, the sighted man has a supersensory view of an aspect of reality. However, the blind are faced with a dilemma. If they accept that the sighted man is not deliberately trying to deceive them, either they simply have to believe, as a matter of blind faith, that the seeing man's incomprehensible visual statements are part of a true description of reality, or they have to dismiss his assertions as insane. It is impossible to describe the perceptions of sight in terms of the other senses.

In any case, as far as the blind man is concerned this discussion would verge on the meaningless. He would perhaps say:

You may be telling the truth, but the rest of us don't even know what you are talking about, let alone how to go about checking your story, so we cannot take it seriously as a scientific hypothesis. We cannot even begin to comprehend what would count as evidence for your idea. All that we are able to verify or falsify is your claim to extrasensory perception, but we are more inclined to believe that you are using telepathy when you demonstrate your extraordinary abilities. But as far as we are concerned, whether or not you are describing reality is a matter of indifference. It may be true in the way that the existence of God or fairies may be true, but it is not scientific. Even if we accept that you are describing reality, the only conceivable way it can be of use to us is if we rely absolutely on you. We would have to have complete faith in your wisdom and powers as a sort of high priest of a realm that only you can access. If we took you seriously, it would be the death of science and the beginning of a new cold age ruled by superstition and insecurity, since the Universe would become incomprehensible to us. We would have to sacrifice our independence and self-reliance for your truths.

The sighted man might reply:

Far from making your Universe more bewildering my greater understanding and knowledge will help you to make more sense of everything. At first it may take a leap of faith for you to take my ideas seriously as a working hypothesis, but this does not mean you will have to suspend your disbelief for long. You can never understand the Universe in terms of the experience of colors, but does this matter? The truth about the way the Universe works transcends the exact way we experience it. Although you can never know what it feels like to apprehend colors, you can in principle give some form of indirect meaning to their existence. Even if you cannot translate the term "red" into tactile equivalents, you can give it an abstract or logical meaning. You can intellectualize it. If you understand that words like "blue," "red," "green," and "yellow" cannot occupy the same logical space, and that a sentence like "This ball is entirely red and entirely blue" is nonsense, then you will have understood the meaning of colors.

As seekers of the truth about the way the objective world is, we have to leave ourselves behind. The way we feel is irrelevant. Once you have understood the essential logical meaning of my statements, it is perfectly possible for you to participate in an investigation of their validity. You can determine whether or not they are logically coherent and consistent with your direct or indirect experientially derived ideas of reality. Provided we are speaking the same logical language, we are talking about the same underlying reality. Once you have understood me I am no longer special or mad, and therefore not a threat to your system. On the contrary, by handing you a coherent and rich theory of the Universe I am not only giving you something interesting to investigate, but am providing you with a few steps on the ladder to transcend the thraldom of your limited senses. I am inviting you to move beyond mere corporeal experience and up into the dispassionate realms of logic, where you can sniff the heady air of essential reality.

The blind man would probably respond in the following way:

What you have said is very convincing, and I agree with much of it. I accept your implicit assumption that it is impossible to separate theory from evidence. There is no such thing as pure experience. The very act of sensing a quality in an object is theory loaded. We have to learn that certain shapes, textures, dimensions, and so on correspond to, but are also separate from, a class of objects. In other words, our apprehension of even a simple object like a child's ball is logically laden. If I were suddenly given eyes, it would be impossible for me to make any sense of these impressions. I would experience a jumble of confusing feelings which would have absolutely no meaning. I would not be able to name even the simplest sensation. We need loads of intellectual baggage for every cognitive act. For the word "ball," or our collection of precepts which we interpret as a ball, to have any meaning we need a pretty comprehensive grasp of language and a correspondingly coherent idea of the world.

It is also true that we can abstract the purely logical aspect of language from its references. We have grammatical rules which we can look at independently of the meaning. We can even devise perfectly comprehensible artificial languages devoid of semantic content. For instance, the purely syntactic metalanguage of mathematics is not only universally understood by all cultures and diverse scientific disciplines, but is so useful that we are inclined to regard these grammatical rules as representative of the rules of nature, and are seduced into the belief that logical structures reflect the natural order of things. Many of us behave as if mathematics is a natural language. One can imagine that mathematics, or even rigidly rule-bound natural languages, would be completely intelligible to intellectual beings who have the ability to communicate but no sensory contact with the external world. To such an acorporeal being, reality would consist in nothing but grammatical or logical rules. He would understand the word "red" only in terms of its logical field. Just as the essential meaning of "pawn" in a game of chess is the moves it can make and its role as part of the structure and dynamics of the game, so the color "red" can be seen as perfectly understood if one knows how to use it within a language. Computers can do that with simple languages like mathematics, even simpler games like chess, and with other logically transparent languages.

It seems to me that you are suggesting that all we need is to behave like a computer, or other comprehensively briefed moron, in order to demonstrate a grasp of transcendental reality. I am not trying to minimize the role of the intellect or the crucial part logic plays in our conception of reality. However, I strongly believe that sensory perceptions and the structure of our minds are totally inseparable. One without the other would result in either amoebas or computers, neither of which can ever begin to comprehend what we think of as the Universe or what exists. Even if computers with no sensory experience could evolve something like intellects, their consciousness and therefore their models of reality would be so different from ours as to be beyond our understanding. They would, in effect, occupy a completely empty Universe where concepts such as existence or nonexistence would have no meaning. There would be no distinction between what is or what might have been or might be. All possibilities would be equally real.

You are inviting us to be like robots with regard to visual reality. You are suggesting that you can, so to speak, program us to simulate a real understanding of your visual world such that we could engage in intelligible discourse about it. We might even be able to fool ourselves that we really understand, and no doubt the more accomplished among us when talking to baffled students would be heard to make remarks like "You just have to forget about your common-sense way of interpreting the world," or, "Reality is not intuitively obvious." There is no way of arguing against such assertions because one has to abandon one's normal rational tools, which we know how to use but which are far too complex and opaque to analyze into a set of rules which we could give to a computer. Therefore one cannot agree or disagree, one can only comply or turn away. The incentive for compliance is the fallacious intimation that refusal is the result of reactionary stupidity. A few may be impressed and wish to be initiated into the restrictive bondage of this perceived intellectual elite, but most of us will retain our mental independence, simply ignore it, and go about making sense of the Universe in terms that we really understand and can believe in. You are the only person who will fully understand or unreservedly believe any of this, but that is only because you can experience it. To you it will be ordinary common sense. Your reality is not and never can be our reality. Anything we have to say about the visual world can never be more than a sort of logical game with transparent rules which may be fun but cannot represent reality.

Truth is not interesting or even meaningful in itself. All that really matters is our relationship with what we think of as the objective world. What we can apprehend, understand, and communicate is everything. Any reality that is inaccessible in principle either through possible direct or indirect experience or our ordinary inductive reasoning is, as far as we are concerned, nonexistent. In other words, any description of the Universe that we cannot in principle verify or falsify by experiment or observation is meaningless. This is just a game played by mathematicians and theologians. While it is true that we cannot access reality through experience or common sense alone, it is equally true that we cannot do so only by logical or mathematical reasoning.

This fictional discussion serves to polarize the positions of rationalism and empiricism, as well as to give a feel for the difference between natural reasoning and formal logic. It is, however, a deceptive oversimplification. It might erroneously suggest that all scientists are conscious of taking philosophical stands, or that they are implicitly exclusively committed to one of these views. All science is a selective mixture of both natural thinking and mathematical physics, and both rationalism and empiricism, with varying degrees of emphasis on each position. The extent to which individual scientists implicitly subscribe to one view or the other depends largely on their role within a particular scientific discipline. For instance, whether or not they are conscious of it, theoretical physicists are far more likely to behave in accordance with the rationalist position, whereas experimental or observational physicists are inclined to the more pragmatic empirical approach.

Einstein interpreted the quantum physicists' philosophical position as a simple denial of the existence of what cannot be observed. Like the blind authorities in Wells's tale, he thought of them as saying that what cannot be apprehended does not exist. Thinking that he was engaging them in a *reductio ad absurdum,* he asked whether they could seriously suggest that the Moon exists only while being observed. In a way this is what their position committed them to, but it is not an absurdity when interpreted correctly. As I indicated previously, it is in fact an aspect of a perfectly valid philosophical position, and it is for this reason that Einstein was defeated in his attempts to refute their argument.

Einstein felt that it was invalid to deny elementary particles a discrete, observer-independent existence. He thought that particle physicists were being irrational and perverse in their refusal to make the perfectly valid assumption that these particles, like any other physical object, have unique identities specific in time and space. This is, after all, implicit in the definition of a physical object. The act of saying "an electron" carries this assumption with it. It is part of the logic of naming. A statement like "This ball has no identity" is a contradiction in terms, since one has already identified the ball in the process of talking about it. One might not know where the ball is, and so be unable to predict where it will be next, but that is not the same as saying that the ball is in itself indefinite, and has either no location or an infinity of locations. All this is just a confusion between the observer's knowledge of the object and the object as it is in itself. It is a type of fallacy arising from the radical skepticism inherent in empiricism. Einstein felt that quantum physicists were going too far, beyond the point where analysis must stop, and were consequently talking the sort of nonsense that arises from asking meaningless questions. We should draw a clear distinction between the objective and subjective, and know when we are talking about external reality or when we are talking about our relationship to this reality. It makes no sense to deny the basic assumption that there is an external reality which is independent of our perceptions. Also, unless one has the fundamental concept of independent identity, it is completely impossible to make sense of anything. The notion of identity is necessary for the survival of even the most primitive sentient being.

Thus it is hardly surprising that Einstein was outraged by the position taken by quantum physicists. He saw them as trying to overturn the most fundamental tenets of scientific reasoning, and indeed ordinary thinking. To him this was intellectual anarchy. It is impossible to have intelligible debate without some agreed premises. Although they must have agreed on the rules of mathematical discourse, this was not a mathematical debate, and therefore Einstein could not demonstrate the fallacy of their position in mathematical terms, although he tried to. It is as pointless as trying to argue whether or not someone could have had the dream they described. This is partly because what is permissible in the self-sufficient and self-referential logic of mathematics is not always allowable in natural thinking, which represents our relationship with the real world and distinguishes our intellect from that of robots.

In the end, although he maintained a running argument with Niels Bohr, Einstein left the field. He believed that quantum physicists, in their rigorous pursuit of uncontaminated certainty, had quarantined themselves in the bizarre realm of radical skepticism, thus putting themselves beyond the reach of normal rational discourse or traditional philosophical challenge. This caused Einstein to turn his back on the realm of quantum physics, which he had helped bring into being and for which he had won a Nobel Prize.

But in spite of Einstein's repudiation of the conceptual foundation of quantum physics, modern science has undoubtedly been molded by the philosophy of logical positivism and its derivatives such as Karl Popper's doctrine of falsifiability or "testability." So it is generally accepted that empiricism and therefore the requirement for objective evidence is what distinguishes science from metaphysics or mysticism. Nevertheless, as we have seen and will see, cosmology is based primarily on the essentially rationalist ideas of theoreticians such as Albert Einstein or Isaac Newton and at times comes very close to the ideas of mysticism. The Universe is almost inaccessible to empirical evidence, so we are heavily dependent on good theories although enough of it is anchored in the work of observational astronomers to lift it above the realm of metaphysics, at least in part.

THE DARK WORLD OF
UNCREATED REALITY

There all the barrel-hoops are knit,
There all the serpent-tails are bit,
There all the gyres converge in one,
There all the planets drop in the Sun.
— W. B. YEATS, *Supernatural Songs IV*, "There"

ISAAC NEWTON ONCE SAID that all his knowledge and understanding of the Universe was no more than playing with stones and shells on the seashore of the vast imponderable ocean of truth. Since then we have pushed back the sea to the most distant quasars so that they now form the seashore of our island within space-time. Everything within the remote horizon of these most distant celestial objects comes within the scope of ordinary astronomy. In big-bang cosmology this empirically accessible realm has been the first observation point, the earliest point in the history of the Universe that we can actually observe. Before the discovery of the microwave background, this delineated the totality of the known Universe lapping on the shores of the vast unknown. Anything beyond the most distant quasars could only be a matter of what Fred Hoyle thinks of as metaphysical speculation, the cosmological theory of the big bang. The detection of the microwave background opened up a new island beacon in the sea of darkness, remote from the limits of our distant quasar horizon.

There is an enormous and enigmatic ocean between our island in time and the event that resulted in the microwave radiation, when matter decoupled from light and the Universe changed from an incandescent opaque cloud to become transparent. The nature of these dark ages between the release of light and the formation of the objects and structures that we see is still obscure, although, as the term "observation point" would imply, dim shapes suggest themselves from the perspective of the remote but observationally solid island that is the fossil evidence of the microwave background. The next empirically established observation point is provided by the fossil remains of the creation of the building blocks of ordinary matter, the epoch of nucleosynthesis.[1] The earliest millionths of a second of creation before this are justifiably thought of as metaphysical speculation, and will remain so until the next primordial relics are found. Without observational evidence, cosmology comes to resemble a theological debate in which appeal to authority and personal conviction decide the truth. The empirically testable ideas of the microwave background and the theory of nucleosynthesis have lifted the big bang above the level of metaphysics by providing islands in the ocean of ignorance which lie beyond the shores of the most remote quasars.

So, ironically, a bitter opponent of the big bang, Fred Hoyle developed the foundations of nucleosynthesis, one of the most important consequences of the big bang, while two telephone engineers made the even more significant observation, the discovery of the microwave background, which, apart from Edwin Hubble's perception that the Universe seems to be expanding, is probably the greatest astronomical discovery of the century. The idea that most of the Universe is composed of primordial black holes may lead to the fourth and most remote observation point in the unfathomed sea of time.

Primordial black holes would form before the light elements such as hydrogen and helium, the ingredients of all the ordinary atomic matter which makes up everything that emits, absorbs, or reflects light and other forms of electromagnetic radiation. The stars, the Moon, and our bodies are all, ultimately, products of nucleosynthesis, whereas primordial black holes and other largely speculative exotic stuff such as "quark nuggets" owe their existence to processes that predate the formation of atoms. We know the nature and abundance of atomic material, and have a good idea of how it came into being. Because we can observe what atomic material exists in the Universe today, nucleosynthesis is an empirically accessible observational theory. By contrast, theories about the very earliest moments of the Universe are extremely speculative.

Our ideas of the history of the Universe since matter decoupled from light are more certain, since we can tell roughly the way things must have been to explain the present structures we see. The fact that the visible Universe is manifestly inhomogeneous, in that we see matter strongly clumped into galaxies and larger structures, means that there must have been small density fluctuations in the early Universe to seed the initial collapse.[2] There must have been very slight condensations to start the process of matter accumulating into galaxies and clusters of galaxies. If the Universe were totally smooth at the time that matter decoupled from light, it would still be in the same state. Nothing would have happened. It would still be the same as it was then, albeit a lot cooler and more attenuated. If, instead, the density fluctuations were too large, then all matter would have rapidly collapsed into superdense objects, and even black holes.

One of the beauties of the theory of inflation is that it predicts the existence of clumpiness quite naturally and uniformly on all length scales at about the right level. It explains both the smoothness on large distance scales and the small ripples needed to form structures.[3] Before the 1980s, primordial ripples were ad hoc assumptions about the initial conditions of the Universe. Without the theory of inflation, the big-bang model could provide no natural way of producing density fluctuations which would serve as the seeds of the structures that we observe. Obviously this inability to explain the slight inhomogeneity necessary for the Universe to have the form that it clearly does was rightly seen by steady-statesmen as an important reason for distrusting the big-bang theory. The fact that the microwave background appeared to be absolutely uniform was another weakness of the big-bang picture of the genesis and development of the Universe. However, since the theory of inflation provided a powerful theoretical underpinning for the existence of slight fluctuations in the cosmic fabric when the Universe became transparent, astronomers were confident that it was just a matter of time before ripples in the microwave background radiation were detected. If the microwave background radiation really is the afterglow of creation, either there are slight fluctuations in its temperature, or the entire big-bang theory including all its variations, such as the theory of inflation, have to be wrong.

In 1992, two radio astronomy groups claimed to have found these ripples in the fossil relic of the early Universe when matter decoupled from light. NASA's Cosmic Background Explorer (COBE) satellite group, headed by George Smoot,[4] had immense resources at its disposal, including hundreds of collaborators and a sophisticated public relations machine, all dedicated to the main objective of finding ripples in the microwave background. The British group, headed by Rod Davies, consists of about a half dozen astronomers based at Jodrell Bank and uses a radio telescope in Tenerife. Although COBE's results are more comprehensive, they are, from a scientific point of view, substantially the same as those obtained by the low-profile British group. They appear to have obtained their results before the Americans but, being British and consequently naturally diffident about their accomplishments, did not publicize their observation that the cosmic background radiation is indeed rippled until after they were scooped by the enormous international fanfare accorded to COBE. Also, they were not under the same pressure to produce and release spectacular results that is typical of high-budget, single-purpose projects such as COBE with its huge concentration of vested interests.

For two independent groups to arrive at substantially the same result under extremely difficult circumstances, using procedures beset by uncertainties at almost every level, is a remarkable achievement. It is hard not to be impressed by these observational triumphs, and I have very little difficulty in believing their results. Nevertheless, it is a little disconcerting that what is predicted by the big bang should so conveniently be confirmed by a powerful observational group.

What if COBE had found that the microwave background is essentially uniform? This would have been a truly revolutionary observation which, if accepted, would have caused a rewriting of all the textbooks. In spite of COBE's huge vested interests and political muscle, it would only reluctantly have been believed, and certainly not as readily as the expected result was embraced and celebrated. However, it is unlikely that such a subversive finding would have been published, let alone publicized, without a lot of soul searching and reworking of the data. The initial assumption would have been that, even if the group had done its analysis correctly, there must be something wrong with their methodology. It is likely that the COBE group would eventually have seen their inability to find fluctuations in the microwave background as an observational defeat rather than a great new discovery overturning many of our assumptions about the nature of the Universe.

For instance, failure to detect the expected fluctuations might have indicated that the microwave background is not, after all, the afterglow of creation, but some other phenomenon such as the black-body radiation produced by Hoyle's needlelike iron molecules, which are the detritus of stellar processes and would emit microwaves at exactly the temperature observed in the cosmic background radiation. Alternatively, it could be a manifestation of the idea, current among dissenters, that light gets "tired" and that consequently the degradation of starlight would produce the same type of homogeneous black-body radiation as the cosmic background radiation without the ripples.[5] Nevertheless, though total uniformity might have weakened the case for a big-bang Universe, since the detection of the microwave background is seen as an important observational confirmation, it would not disprove it.

The fact that ripples of about the right size have been detected by two independent groups is a very powerful endorsement of the microwave background as the afterglow of creation within the big-bang model of the Universe. Dissenters would be extremely hard put to challenge the validity of these observations, although they could question their importance since they really add nothing new to our understanding of the Universe. Although the detection of the ripples is undoubtedly an observational triumph, it is hardly an earthshattering discovery. Since astronomers have always made the assumption that the microwave background is inhomogeneous, the discovery that this is indeed the case does not rewrite the textbooks, but merely provides a quantitative footnote. The tremendous fuss surrounding these observations, even to the extent of our supreme cosmologist Stephen Hawking heralding COBE's results as "the scientific discovery of the century, if not of all time,"[6] could fuel Fred Hoyle's seditious fire. Hoyle might claim that this confirms his contention that modern cosmology operates on the principle of maximum trivialization, and that the more banal and uncontroversial an idea or discovery is, the greater are its chances of being celebrated as important.

So, as we always suspected, the afterglow of creation displays the slight density fluctuations that have to be there in order to explain the structures that we can see. What is more, these slight condensations of matter had already to be in place when matter decoupled from radiation and the Universe became transparent. If matter started to condense only after the Universe became transparent, there would not be enough time for it to have evolved the structure it now displays. But it was impossible for atomic matter to have condensed to any extent before decoupling because, being susceptible to radiation, the instant such "baryonic" particles gravitated toward each other in the primordial soup, they would have been blasted apart by radiation. If it was impossible for atomic matter to condense in the extreme conditions of the early Universe, how do we explain the fact that matter condensation nevertheless took place, except by an appeal to the existence of forms of matter which are not affected by radiation?

Hence there is a further problem whose solution lies in the primordial existence of dark matter which is nonatomic, or "nonbaryonic," and so unsusceptible to radiation. Thus, in addition to the need for sufficient mass to enable the Universe to condense into structures, there is another reason for the belief that there is a significant amount of, not merely so-called missing mass, but completely invisible primordial material. Unperturbed by the blandishments of electromagnetic radiation, such stuff would have been able to condense before matter decoupled from light, so it would have seeded the formation of all objects and structures. Thus nonbaryonic matter could have begun to determine the shape of the Universe well before ordinary matter was released from the disruptive thrall of radiation, and in its turn enabled to participate in the gravitationally attractive business of forming compact bodies, galaxies, clusters, and the large-scale irregular honeycomb structures into which most of the visible Universe seems to be organized.

For this reason alone, it is believed that primordial dark matter must permeate the Universe. But the exact nature of this nonatomic stuff has been purely speculative and, until my case for a population of ubiquitous primordial black holes, there has not been (in the words of Fred Hoyle) "a scintilla of evidence for the existence" of any such entities. This is in spite of a vigorous search for nonbaryonic elementary particles. So a number of fundamental questions are still being asked. Is nonatomic dark matter in the form of so-called hot dark matter or cold dark matter? In other words, is it mainly very fast moving stuff like neutrinos, or slow nonbaryonic entities? If it is cold, does it occur as compact bodies such as primordial black holes, or as elementary particles? (Hot dark matter cannot be compact, so must occur as elementary particles.) There is a panoply of theories in favor of various types of elementary particles for which particle physicists and other cosmologists are pursuing observational evidence. Such evidence will be extremely difficult to find, mainly because interactions of these particles with the atomic matter of detectors would be almost unimaginably rare, even if they exist at all. However, cosmologists seem to be placing their main reliance on computer simulations to help settle the question of whether we live in a cold dark matter, hot dark matter, or mixed dark matter Universe. In my opinion this might be a mistake. Computers are not objective but have to work within our assumptions and prejudices. Computer simulations cannot tell us anything new about the Universe. They are just a way of clarifying the arguments contained in our theories and speculations. They help us to illustrate and understand our ideas better, but nothing can be settled in this way.

Knowing the exact nature of the dark matter would give us an insight into the evolution of the structure of the Universe, thus illuminating the dark ages between the distant events that caused the microwave background radiation and the shapes we now see delineated by the sprinkling of visible atomic matter and bounded by the most remote quasars. But perhaps the most important reason for trying to determine the identity of dark matter is to provide a new island beacon in the very earliest Universe, before the synthesis of atoms and well before the Universe became transparent, thus allowing us a glimpse into the dark primordial world that existed before what we might think of as the true creation, the coming into existence of the stuff that makes up everything we know and in which we have our being. The identity of primordial dark matter not only describes the substance of the material Universe, but, as a relic of the state of the Universe in its very earliest moments gives us an important observational clue to the process of creation, thus helping to lift it above the realm of metaphysical speculation. It brings necessarily rationalist theories about creation within the reach of empirical science. But conversely, observations that bear on the virtually inaccessible realm of the beginning of everything are highly dependent on good theories.

However, contrary to Einstein's Popperian belief that "A theory can be proved by experiment; but no path leads from experiment to the birth of a theory,"[7] the relationship is far more symmetrical in practice. Theory and evidence are interdependent, especially in cosmology. Although theoreticians often go about their business with very little regard for the work of observers and, as illustrated by the attitude of Gerard De Vaucouleurs, observers frequently display a haughty disregard for rationalist theories, real progress has only been made as a consequence of the interplay between theoreticians and observers. For example, without Einstein's theory of relativity and the work of the Jesuit priest Georges Lemaître, Edwin Hubble's observation that distant galaxies are receding would have remained a mystifying oddity instead of forming the basis of a revolutionary new cosmology, the big-bang theory. Equally, without observers such as Hubble, Einstein's ideas would have remained an abstract mathematical tour de force, but would not have taken on the enormous cosmological significance that they now have. Madame Einstein is reputed to have said to Hubble when marveling at the huge amount of technology and staff at his disposal, "My husband does the same thing on the back of old envelopes." But this misses the point. Einstein and Hubble were doing different things. Their respective greatness and the emergence of the big-bang theory is not a result of duplication, but of symbiosis between rationalism and empiricism, between theory and evidence. This example runs completely counter to Karl Popper's ideas of the scientific method and is, in my opinion, an illustration of the way that we actually arrive at our scientific beliefs, especially in cosmology. Hubble was no more setting out to test Einstein's ideas when he found that the Universe is expanding than Einstein was trying to explain Hubble's observations when he devised his theory of relativity.

On the face of it, however, Popper's dicta seem to apply to the theory of inflation especially since cosmologists like Fred Hoyle and David Schramm applaud its strength on the basis of making specific testable predictions. Schramm maintains that

All theories of inflation make a dramatic prediction by which they can be readily tested. In the process of solving the age or flatness problem, inflation requires that Ω be almost exactly equal to 1 [the critical density]. No other physical theory has been able to specify its value so precisely and unconditionally. In the past, cosmologists had to rely primarily on guesswork and intuition. Now there is a specific, testable prediction.[8]

He goes on to say that

Experimental proof of inflation—that Ω is very close to 1—seems to be almost at hand. If and when it comes, we will have to confront the likelihood that over 90 percent of the Universe is totally dark and not made of ordinary matter.[9]

In fact the situation is a little more complicated. As well as predicting that Ω is exactly equal to 1, inflation also explains why it is as close to 1 (in the astronomical scale of things) as we already know it to be. Thus inflation provides both a prediction and an explanation for the value of omega.

Bill Press and Jim Gunn demonstrated that if the Universe has the critical density in compact bodies, then every line of sight will be gravitationally lensed.[10] The converse, that if every line of sight is lensed, then the Universe has the critical density, is also suggested by their work. But exactly how close Ω must be to unity is not clear. However, on this basis, as we shall see, my evidence that every line of sight is indeed gravitationally lensed may be seen as a confirmation of the theory of inflation.

But I suspect that this is not the type of "proof" that Schramm had in mind. It seems that, in line with Popper's picture of the scientific process, Schramm visualizes the relationship between theory and evidence as being more like the deliberate way in which the COBE team set out to confirm the prediction that the microwave background manifests temperature fluctuations; that in a way, evidence follows theory. It does happen this way, but usually contributes nothing of far-reaching significance, and it often simply makes us suspicious when observers so conveniently find what they looking for as seems to have been the case with Eddington.

I most certainly did not set out to test the theory of inflation, nor was I particularly concerned with finding dark matter. All that interested me was examining the nature of quasars, which seemed unlikely to contribute very much to our understanding of the Universe. Although primordial black holes had been mentioned as possible contenders for the missing mass and as cosmic seeds, and although in 1982 Crawford and Schramm pointed out that under favorable circumstances black holes with the mass of the planet Jupiter could be formed when quarks started to condense into protons and neutrons,[11] nobody saw this as a serious possibility. There was no specific theoretical prediction that primordial black holes existed in any abundance, and little attempt to search for such compact dark objects. Therefore the implications of my discovery came as a complete surprise, both to me and to the rest of the astronomical community. Because it did not conform to the standard Popperian model of the scientific process, where the function of observational evidence is to confirm specific predictions of theories, there was a tendency to regard my conclusions with suspicion, especially since they conflicted with the received wisdom about the nature of quasar variability.

On the other hand, there were, and are, a plethora of theories predicting exotic elementary particles of one form or another, some of which would be massive enough to be dark matter candidates. There are a number of high budget and high profile projects dedicated to the search for the bulk of the Universe in the form of such particles, which include "axions" and "WIMPs" (weakly interacting massive particles), as well as attempts to determine whether or not neutrinos have any mass. The astronomical community would be very pleased and excited, but not surprised, if any such stuff was discovered, since that would be broadly in line with theoretical expectations. It is nevertheless worth remembering that the predictions of these theories are only of the vaguest nature.[12]

Although an important reason for the admiration accorded to the theory of inflation by big-bangers and even by some dissenters is because it makes a specific testable prediction, the main reason for its widespread acceptability is, quite simply, because the big bang cannot survive without something like inflation. As David Schramm says,

> In the decade since it was first proposed, the theory of inflation has
> become one of the dominant ideas in all of big-bang cosmology.

Though still a hypothesis, it is a very compelling one that many cosmologists already take for granted. It has become increasingly difficult to imagine how our Universe could have begun without invoking something like inflation to clear away the monopoles and make it end up so flat, no matter what it was like before.[13]

Strictly speaking, this description applies only to so-called new inflation, a revision of Guth's original model proposed by a number of cosmologists in the early 1980s. There was a major problem associated with Guth's original idea. It had no mechanism for bringing everything that was blasted apart by inflation back into the same event horizon. As the Universe condensed from its initial state, in which all forces were unified in a single superforce (as described by grand unified theories, or GUTs),[14] resulting cracks and bubbles in the fabric of space-time would have been blasted apart from each other by inflation, never to meet again. Another way of visualizing this is to think of inflation as constantly smoothing out any ripples so that they are obliterated the instant they appear, with the result that the Universe could never complete its attempt to condense from the original state of undifferentiated symmetry where all forces are unified in a superforce. Analogously, the homogeneity and symmetry of a body of water as it cooled could never change into the asymmetrical flawed structures of ice if some mechanism were constantly blowing the water particles apart from each other. Perhaps a better analogy would be the idea of the conversion of a quantity of soapy water into bubbles where the individual bubbles are blown away from each other at such a rate that they could never coalesce into a single body of froth where the individual bubbles expand and merge into one another, which is roughly how Guth saw the Universe.

The inflation revisionists had a neat solution. They removed the requirement that the entire GUT-era Universe coalesce into a single Universe incorporating all the bubbles and fractures of its transition into a perturbed primordial froth. They proposed that the entire observable Universe originated inside a *single* bubble, and that all that went before the inflationary era had sped off into independent bubble universes. In other words, our Universe is bounded by an event horizon or crack in space-time, and does not contain such flaws within itself. In the new inflation, entities such as cosmic strings, the manifestation of such space-time fractures, cannot exist in the observable Universe. Stephen Hawking's creation model is very close to new inflation, differing only in that nothing predates the era of inflation. According to him, inflation is identical to the big bang. In many ways this is a more elegant picture since it removes the need for a myriad of other universes, totally inaccessible and verging on the metaphysical.

Guth came up with the idea of inflation as an explanation for the absence of extremely massive particles known as magnetic monopoles, which according to the big-bang model without inflation would have had to exist in such abundance that their combined mass would long ago have caused the Universe to implode into a black hole. For each event horizon, or space-time bubble, formed in the GUT epoch, a magnetic monopole had to be produced. This is comparable to the tiny excess of atomic matter, the baryons, which had to be produced in this breakdown of the GUT. In the big bang without inflation, it takes about 10^{80} horizon volumes, bubbles of space-time isolated within their own event horizons, to create enough atomic matter to make up our present Universe, so there would have to be 10^{80} magnetic monopoles, about the same as the number of baryons. Given that a monopole is over 1 million billion (10^{15}) times more massive than a proton, which is the most familiar type of baryon, this would mean that there would have to be about 1 million billion times more mass in the Universe than is accounted for by ordinary atomic matter. The problem was to find a mechanism for getting rid of the vast majority of monopoles without also similarly reducing the amount of baryons.

Guth found a solution in 1979 which he published in 1981. In the process he solved a number of other problems associated with the big bang. He is quoted as saying, "I discovered that the Universe grew exponentially, inflating like a balloon. I felt it was a spectacular realization." Magnetic monopoles that formed at the beginning of the era would have become so attenuated by the exponential explosion of the Universe that there would only be one monopole within the observable Universe. Because, in a manner of speaking, the Universe would have expanded faster than the speed of light, and hence faster than the transmission of any information, different parts would become causally disconnected, thus effectively isolating each monopole

112

within its own event horizon. In Guth's model, the event horizons formed in the GUT epoch are separated only by the inability to communicate with one another. In time, these horizons start to overlap as the passage of light, so to speak, catches up with and overtakes the Universe's expansion rate. This idea is fundamentally different from the idea of new inflation, where these horizon volumes are separate not merely because light has not yet had enough time to travel from one to the other, but because they are completely and irrevocably isolated from one another by the unsurpassable physical boundaries of their event horizons and the realm of the unified superforce, which still holds sway in the interstices. Thus in new inflation the horizon volumes formed at the breakdown of the GUT epoch are effectively totally separate universes with absolutely no possibility ever of any form of communication between them. As far as we are concerned they are as good as nonexistent, which is one reason why Stephen Hawking's model seems more sensible. If they are in effect nonexistent then why not settle for a model that does not require them?[15]

But what of the building blocks of atomic matter, the baryons? Why would they not, like the monopoles, be attenuated by inflation to the extent that there would be only one baryon in the observable Universe? The answer to this is very simple. They formed after the era of inflation, so they would not have been diluted almost out of existence, leaving a virtually empty Universe. Changes from one state to another, such as the featureless homogeneity of water turning into the lattice structure of ice with its cracks and bubbles, are known as phase transitions. Such phase transitions would have taken place as the Universe cooled, creating the environment for the formation of new entities such as primordial black holes and baryons. According to both Guth's early version and new inflation, the phase transitions that occurred during and after inflation would be part of our Universe, so anything produced by them should still be present as relics.

Although physicists have yet to produce a satisfactory model for the quantum processes that may have occurred during inflation, the structure of the Universe we see today could have had its ultimate origin in the minute inhomogeneities caused by random oscillations due to quantum fluctuations in the energy fields of the inflationary era. So, rather than representing the phase transition of the breakdown of the symmetry of the superforce at the end of the GUT era, the structure of the Universe probably had its genesis in the slight quiverings of space as it was created during inflation.

Thus inflation explains not only why the Universe exists at all, but also why it is smooth enough not to have produced an overabundance of large structures, which would have devastated the Universe, yet flawed enough to account for the structures that we observe. The nearly homogeneous nature of the microwave background is accounted for, as is the very slight clumpiness observed by COBE and the Jodrell Bank team. Inflation also explains why, on the large scale, the Universe is the same in all directions, even though different parts of it cannot be causally connected, since light would not have had time to travel from, say, the farthest reaches of the observable Universe in the north to those of the south.

The big-bang model without inflation could not explain how, given a certain amount of turbulence in the early Universe, areas constrained by different event horizons would nevertheless have maintained their uniformity with one another. The inflation mechanism solves this mystery simply by ironing out any irregularities that might occur. No matter how turbulent or how irregular any region might have been, it would be forced back into smoothness, and hence conformity with other regions. Thus the Universe will remain flat, which is another way of saying that it will maintain the critical density. This incidentally explains why the Universe today is, in cosmological terms, so close to the critical density, at least within a factor of 10. For this to be the case, the early Universe had to have almost exactly the critical density since even the tiniest departure from this state of affairs would have been magnified to the extent that the Universe would long ago either have disappeared into a black hole, or blasted away into infinity. It could not have got as old as 15 billion years and be as it is today unless there was a critical balance. But the visible Universe contributes less than 1 percent of the critical density, so in an inflationary Universe, in what form is the rest of the matter? Does it mean that the vast bulk of the material Universe is invisible? If we can establish the existence of this dark matter, and show that it has the critical density, we will have confirmed the major prediction of the theory of inflation.

Recalling all the issues surrounding the measure of the Hubble constant, I maintained that the current high estimate of this parameter, about 80 km/s/Mpc, should be treated with suspicion, if for no other reason than that it conflicts with the theory of inflation, and that this crucial modification to the big-bang theory had to be right for the survival of our current idea of the Universe. To my mind, falsifying the theory of inflation is tantamount to falsifying the entire big-bang model. Conversely, verifying the theory of inflation has the effect of establishing its validity.

Having said all this, there are more important reasons for our belief that most of the Universe exists as dark matter. The search for the identity of the vast bulk of the Universe has been motivated primarily by observational rather than theoretical considerations. Nevertheless the fact that the theory of inflation unequivocally requires that the Universe should have the critical density has certainly added extra impetus to the search for the missing mass.

9

DETRITUS

Mind in its purest play is like some bat
That beats about in caverns all alone,
Contriving by a kind of senseless wit
Not to conclude against a wall of stone.

It has no need to falter or explore;
Darkly it knows what obstacles are there,
And so may weave and flitter, dip and soar
In perfect courses through the blackest air.

And has this simile a like perfection?
The mind is like a bat. Precisely. Save
That in the very happiest intellection
A graceful error may correct the cave.

—RICHARD WILBUR, *"Mind"*

W E HAVE SEEN THAT IT IS A CONSEQUENCE of modern big-bang theory which includes inflation that at least 95 percent of the Universe is made up of dark matter, but there are also observational reasons for believing that most of the mass of the Universe is invisible. Only a small proportion of this material can be the ordinary atomic matter that makes up objects such as our bodies and the stars. It is only this atomic material which we can observe with telescopes, our eyes, and other radiation detectors, which are of course themselves atomic.

Our present understanding of the big bang and the early evolution of the Universe places tight limits on the amount of ordinary atomic material. The theory of nucleosynthesis describes how, soon after the birth of the Universe, such primordial atomic matter was formed, light elements such as hydrogen and helium providing the building blocks for the subsequent creation in stars of heavier atoms such as oxygen. Any nonatomic primordial matter would represent the stuff that either failed or never had the potential to condense into primordial atoms before and during this period of nucleosynthesis. In the 1970s, theories of nucleosynthesis and observations of the abundances of atomic elements such as helium and deuterium put an upper limit on the amount of atomic matter: it could contribute no more than 10 percent of the critical density.[1] Thus nonatomic primordial matter has represented the huge bulk of the material state of the Universe almost from the instant of its genesis. Such stuff cannot interact with any form of electromagnetic radiation such as gamma rays, X-rays, visible light, microwaves, radio waves, or whatever. Nonatomic dark matter does not absorb, block, emit, or reflect any such rays or waves. It is not affected by nor does it influence electromagnetic radiation, and therefore cannot be observed by any normal type of telescope or radiation detector. So how can observers even speculate that the vast bulk of our Universe is composed of such profoundly invisible matter, let alone go about finding this exotic, almost supernatural stuff whose tenuous claim to existence is just its pervasive gravitational influence on the visible Universe of atoms?

The problem of dark matter was originally an issue raised by observational astronomy, but the general belief is that the solution is not to be found in this domain, and probably not even in conventional theoretical astrophysics, but in the realm of particle physics. The most popular candidates for dark matter are nonatomic primordial particles, which would have only rare and fleeting contact with the atomic material with which we are familiar. Thus it can only be detected by observing almost unimaginably rare interactions with the elementary particles that form atoms.

Massive bodies (those that have *any* mass) move relative to one another according to well-defined gravitational principles. All matter is attractive; the more massive an object, the greater its attraction. Thus the Earth orbits the Sun because it is in a sense continually falling into the Sun. Similarly, the Moon's orbit represents a continual fall toward the Earth. Because the Sun accounts for about 99.9 percent of the mass of our Solar System, it essentially *is* the Solar System. The planets are almost negligible. Their gravitational influence is therefore irrelevant in the context of weighing the mass of the Solar System, although their behavior is an essential means of finding out what this mass is. Their orbital velocities and distances from the Sun serve as measures of the mass of the Solar System, just as the orbital dynamics of the Moon tells us the Earth's mass. However, the calculations are a lot more complex when we measure star systems such as galaxies, where the relevant mass is the combined mass of billions of stars moving under their mutual gravitation, each one influencing and being influenced by all the others.

In simple terms, the more massive a system, the faster bodies such as stars move about within that system. It is relatively easy to assess the mass of a spiral galaxy because the whole galaxy is rotating in a coherent way. All we need to measure is the rate of its rotation relative to its radius. With amorphous or elliptical galaxies this measurement is a bit more complicated and uncertain since the motions of the stars are more chaotic. We have to ascertain the spread of velocities of a representative cross-section of stars.

When weighing a cluster of galaxies, we use basically the same method as when measuring the size of individual galaxies, except that we treat the galaxies within the cluster as individual bodies. Since there are usually a manageable number of them within a system, it is possible to take the entire population into account, rather than just a sample. However, we have to assume that these clusters are in equilibrium and not in a state of expansion or contraction.

Despite these drawbacks, and the consequent disagreement between astronomers about how massive galaxy clusters are, it is at least clear that the study of their kinematics reveals far more mass than in the constituent galaxies. There is now some consensus that the most reliable statistical measurements of mass from the dynamics of individual galaxies are probably those which indicate a hidden mass of about 10 times the visible mass. This falls far short of the measurements from the dynamics of galaxy clusters. These suggest that there is up to 100 times more mass than can be observed in the ordinary way. The large discrepancy between these estimates of hidden mass probably reflects the fact that most of the invisible Universe lies outside visible galaxies.

Thus astronomers have had little idea where to look for most of the mass of the Universe, let alone what form it takes. However, we can safely assume that a great deal of matter is missing, that some of this dark matter lies within visible galaxies and also that some of it is atomic. Observations made using X-ray space telescopes, for instance, seem to rule out most noncompact forms of atomic dark matter, such as clouds of hot gas, as contributing significantly to the total of dark matter. There is, however, the possibility that it is in the form of very cold molecular gas, which at present would be almost impossible to detect. There is also the popular idea that atomic dark matter escapes our attention simply because it comes in the shape of very faint compact bodies such as barely luminous failed stars, and the dimly glowing embers of small burnt out stars, the elusive brown dwarfs and white dwarfs. The search for this interesting atomic component of dark matter is the preoccupation of some observational astronomers (including myself), who have yet to find a significant population of such "massive compact halo objects" or MACHOṣ.

As I have indicated, within our mainstream cosmological models it seems likely that only a very small proportion of dark matter is atomic. The amount of matter that we have already observed is fairly close to the quantity of atoms that could have been synthesized in the big bang. Initially it was because of these theoretical constraints that the larger observational estimates of hidden mass were often simply not believed. Big-bang theoreticians maintained that, since it is impossible for there to be more matter in the Universe than could have been synthesized, observers were making a mistake when they suggested that up to 99 percent of the Universe is hidden. It was only when physicists postulated the existence of primordial nonatomic matter that this discovery by observational astronomers was not only given credence, but became an intriguing and challenging idea at the cutting edge of mainstream cosmology. By the same token, the heart of the dark matter problem was lifted out of the realm of ordinary observational astronomy. Observers were left to deal with the residue, the small amount of atomic dark matter, the identification of which would solve little of significance to the large-scale structure of the Universe. We have nevertheless enthusiastically pursued these elusive MACHOs with extensive observing programs. They have attracted media attention on the basis that, for instance, the supposed sighting of a brown dwarf or whatever is of real significance to the missing

120

mass problem. For example, in 1989 I basked in the brief glory of newspaper and television publicity when I reported the discovery of what appeared to be a population of brown dwarf stars. More recently, in 1993, the detection of low-mass stars in the halo of our Galaxy caused great excitement worldwide.[2] But however fascinating these results may be, they have little bearing on the missing mass problem.

The principal reason for the dark matter problem assuming a far greater significance than astronomers at first believed, and which many still refuse to believe, has to do with the fact that the theory of inflation requires 100 times more matter in the Universe than we have observed. As we have seen, according to inflation, the density of the Universe has to be exactly equal to the critical density. This would mean that the Universe will neither expand to infinity nor start to collapse upon itself, but will maintain a sort of equilibrium between these two states indefinitely. There has to be just enough matter, and hence just enough gravitational attraction, to hold the Universe together without causing it to collapse upon itself. If there were any less mass, the Universe would expand forever into the cold wastes of infinity. If the Universe were much more massive it would long ago have collapsed under its own weight into a singularity, the annihilating furnace of an enormously dense but infinitely small black hole. Either way, the concepts of space and time, and therefore existence, would lose their meaning.

At present there are no theories and no evidence to support the possibility that the mean density of the Universe is greater than the critical density. Accordingly, few astronomers believe that the Universe will eventually collapse. Thus only two possibilities are currently taken seriously: either the Universe will expand to infinity, or it will reach an equilibrium.

As we have seen, the density of the Universe may be expressed in terms of the parameter omega. For an empty Universe $\Omega = 0$, while for a closed Universe Ω is greater than 1. A Universe that tends neither to infinity nor to a singularity has the critical density, $\Omega = 1$. Even if one pushes the data to the most optimistic limits, the total of atomic matter that could have been synthesized in the big bang corresponds to a value no greater than $\Omega = 0.1$, and obviously this total cannot be less than the totality of the visible stuff of the Universe, which is more or less equivalent to $\Omega = 0.01$. Accordingly, there could be as much as 10 times more atomic matter than is discernible. Thus if $\Omega = 0.1$, there may be no need to invoke nonatomic dark matter. The pursuit of MACHOs would assume greater importance at the expense of the relevance of WIMPs, weakly interacting massive particles, which, along with massive neutrinos, are the main dark matter candidates of particle physics. The conclusion must be that the greater the dark matter problem, the less astronomy can contribute to its solution. So, evidently, a low value of Ω enhances the relevance of the work of those astronomers who stick to their belief in a low-mass Universe, regardless of compelling reasons to the contrary. It is in their interest to maintain that the dark matter problem can be resolved entirely by looking for atomic compact entities such as brown dwarfs, non-primordial black holes, and other such elusive bodies.

It is for reasons like this that the theory of inflation is considered problematic. Hence some astronomers are skeptical about inflation, despite the fact that it is necessary to provide solutions to serious difficulties associated with the big-bang theory. Such skeptics must disregard the fact that calculations from the analysis of large galaxy surveys suggest a mass that is consistent with Ω being close to the critical density, and certainly larger than 0.1. The extent of the missing mass problem therefore remains controversial despite the compelling elegance of the theory of Inflation.

However much dark matter there might be, and whatever proportion is atomic, there is certainly a great deal of it. Most astronomers and particle physicists believe that dark matter is in the form of atomic compact bodies, or elementary particles, or both. The task of finding dark matter is therefore in the province of both particle physics and, to a lesser extent, astronomy. The general perception is that there are basically only two candidates for dark matter. These are exotic elementary particles that were synthesized in the unimaginable blast furnace of the big bang, and compact bodies of ordinary stuff, such as the barely luminous failed stars known as brown dwarfs. Thus quests for the missing mass of the Universe have mainly been confined to searches by astronomers for MACHOs within the halo of our Galaxy, and to attempts by particle physicists to detect the speculative WIMPs by esoteric processes using strange detectors such as chlorine lakes in underground caverns.

Some MACHOs can, in principle, be detected by telescopes that are sufficiently sensitive to very low luminosities. Improvements in telescope technology and observation techniques have finally revealed evidence for objects like the dimly glowing brown dwarfs. But no matter how advanced our telescopes become, we could never directly detect dark or invisible objects like black holes by such traditional methods. There is, however, still the dying hope that because of the possible phenomenon of "Hawking radiation,"[3] minute black holes, if they existed, would betray their presence by emitting gamma rays as they evaporate.

Black holes are theoretical entities that represent the ultimate state of collapsed matter. In principle they may be of any mass, but there are certain mass ranges where we believe conditions can easily arise for them to form. For example, dead stars with more than four times the mass of the Sun no longer have the energy to resist the power of their own gravity to crush them almost out of existence, into the secluded annihilation of a singularity. This is the infinitely dense and infinitely small state of being where atoms are decimated into their primordial parts and cannot interact with any form of radiation. Absolutely nothing, not even light or other forms of radiation, can escape the intense downforce of such gravitational whirlpools. Such entities can be thought of as nonatomic compact bodies because they are no longer atomic. Nevertheless, their mass is part of the totality of what came into being as atomic material in the early Universe. The fact that they may have subsequently reverted to a nonatomic state is irrelevant to the business of finding dark matter. Hence nonprimordial black holes are regarded as contributing to the sum of ordinary atomic dark matter, and are therefore included in the list of MACHO candidates that astronomers are hunting down.

Astronomers prefer to express the idea of black holes in terms more consistent with Einstein's theory of general relativity where gravity is almost synonymous with the shape of space. In Einstein's theory, gravity is portrayed not as an attractive force, but as the slopes and curves of space. Gravity is the way in which massive bodies distort space, rather in the way that a heavy ball might stretch and curve the surface of a rubber sheet, causing lighter balls to roll toward it. A black hole is where space is so curved in upon itself that even light and other carriers of information cannot escape, but keep going round and round within the black hole's "event horizon" thinking that they are traveling in a straight line. Like a bottomless whirlpool, a black hole can draw in material and information from the rest of the Universe, but it gives back nothing.

According to this way of thinking, the whole Universe could be finite and similar to a black hole in that it is limited by its own event horizon, within which light can be thought of as traveling in a straight line, whether or not in some absolute sense it curves back upon itself. We define a straight line as the route that light travels, so in this sort of Universe parallel straight lines will eventually meet rather in the way that lines of longitude meet at the poles. In other words, from a detached perspective the overall shape of the Universe may be analogous to the surface of a sphere. In that case, the value of Ω would be greater than 1. The Universe would eventually collapse under its own weight into a singularity when it no longer had the energy to withstand the force of its own gravity. A closed universe, which is also a finite universe, has been pictured as a serpent swallowing its own tail. Its geometry is analogous to that of the surface of a sphere, where the sum of the angles of a triangle is greater than 180 degrees. In an open universe, which is unbounded and infinite, the average density is less than the critical density. It can be thought of as hyperbolic or saddle shaped such that the sum of the angles of a triangle will be less than 180 degrees. Here parallel straight lines as defined by the path light travels will become increasingly remote from and beyond communication with each other in the vastness of infinity.

A flat universe is one that has the critical density, where our idea of straight lines prevails. It represents a type of infinity in that a flat universe in equilibrium will retain its identity for all eternity. Although we think of it as neither open nor closed, it can be regarded as a special case of an open universe since it is analogous to an infinitely extended flat surface. Such a universe has the familiar flat Euclidean geometry, where the angles of a triangle add up to 180 degrees. Here, parallel straight lines will always remain the same distance apart. They are analogous to lines of longitude on a flat Earth, where there is no general surface curvature determining their paths, though hills and valleys will cause local distortions. Thus a two-dimensional map, or an aerial view of a flat Earth, will suggest that the distance between lines of latitude or longitude drawn to the same scale will always be identical. On the ground, however, these distances would vary according to local perturbations such as the extra bulk of Mount Everest which, so to speak, stretches the Earth's surface, thus effectively increasing the distance between lines of longitude or latitude which a land traveler actually has to cover. Although it is perhaps dangerous to take this analogy too far, sea level could be regarded as the two-dimensional state of zero mass, and therefore zero gravity and hence no distortion of space. The third dimension of height of the Earth's surface above sea level would then represent mass. The higher a point on the land, the more it distorts the Earth's surface and therefore the greater its mass and gravitational attraction. In this picture, if a very slender radio mast built at sea level is as high as the summit of Mount Everest, then the mast and the mountain are equally massive since it takes the same amount of extra distance to surmount and descend both objects. They both intrude into the third dimension to the same extent. However, their slopes are very different: the radio mast is more compact. In other words, the greater steepness of the radio mast, in this analogy, represents its greater density. Thus a Jupiter-mass black hole can be thought of as the equivalent of a radio mast in relation to the Mount Everest of Jupiter itself. Jupiter has a diameter 10 times the Earth's, while a Jupiter-mass black hole is about the size of a double bed.

This analogy to some extent sums up Einstein's way of regarding the Universe. It does not commit us to a choice between closed, open, or flat and, whatever the overall situation, there will always be local distortions of space. We cannot step outside the Universe and view the equivalent of lines of latitude and longitude following curved paths, as we can for the Earth. We are trapped within our own valleys and hills within the fluctuating distortions of space itself. For us there cannot be any absolute point of reference.

If we imagine two hikers who both agree to walk in parallel at the same constant speed, but without realizing it, one sticks to level ground while the other takes to the hills, then obviously the hill walker will seem to be covering less ground than the level treader. If they walk in sight of each other, the hill walker will think that he is moving more slowly than the level hiker, who will have the same impression. Each will think that the other has broken their agreement. This is somewhat analogous to the idea of time being relative, though it fails as an analogy in that it represents an asymmetrical relationship, since two space travelers would both see each other as moving more slowly. Each walker has his own frame of reference, which is dictated by the shape of the space he has to traverse. The hill walker thinks he is moving in a straight line because he is unaware of the third dimension of height, which in actuality curves this line and increases the real distance covered. Similarly, a space traveler will appear to an earthbound observer to be traveling at a different rate, and vice versa. Your frame of reference depends on where you are in the Universe and on the shape of your local space.

The curvature of space can be thought of as occurring within a fourth spatial dimension, just as the curvature of the Earth's surface is regarded as occurring in the third dimension. We see the Earth's surface as a two-dimensional aspect of a three-dimensional sphere. By analogy, we can regard the shape of space as the three-dimensional surface of a four-dimensional sphere. But this is pure analogy, since to us a four-dimensional sphere is unimaginable. Because physicists define a straight line as the path light would follow, we think of space itself as being curved. It is just another way of saying that our normal idea of straight lines does not exist in nature, and that the behavior of light and therefore of space is better expressed in terms of the mathematics of curved surfaces. This conceptual breakthrough is on a par with the realization that the Earth's surface is more accurately described it terms of the geometry of the surface of a sphere. However, it would be unnecessarily cumbersome to take the Earth's curvature into account on a local scale, when considering for example the area of a small plot of land. Similarly, for many purposes we find it more convenient to describe the Universe in terms of Newton's ideas. Also, the Earth really is a sphere whereas the idea of space itself being curved is a metaphor. "Curved space" is a catchall expression embodying a complicated set of procedures for mapping the large-scale structure and history of the Universe.

The reason why physicists prefer to think of space itself as curved is because the behavior of light is fundamentally allied to the concepts of space and time, and the geometry of light rays is a curved geometry. For instance, when we say that a galaxy is 1 million light-years away we are also saying that we see it as it was 1 million years ago. Thus when we look out into the Universe we are always looking back in time. If light moved at an infinite speed, we would not need to combine the notions of space and time, nor would we be able to do so. By the same token, we would not be able to say very much about the history of the Universe because everything would be present in the here and now. Thus, in a way, time is inseparable from distance, and therefore from space; one without the other is a useless abstraction. Hence the finite, measurable speed of light gives us some idea of time and space. It therefore determines our concept of the history of the Universe, which would otherwise be no more than a subject for abstract metaphysical debate rather than scientific activity.

Nothing can move faster than the speed of light, and nothing can take a shorter route between two points. In astronomical terms the route that light takes is the standard of straightness, no matter how much it might in some absolute sense curve. It is the means by which we measure the Universe and everything within it. By definition, light cannot deviate from the most direct route possible. We cannot abandon this idea because there is no other standard of measurement. So, in a way, it really is meaningless to think of lines in space as anything but the curved trajectory of the passage of light. Also, since these paths are determined by mass, it makes sense to think of space as being shaped by matter. Just as time can have no meaning without events, so the concept of completely empty space is meaningless. It has the logical and therefore the ontological status of a transitive verb such as "to know," which defines the relationship between the knower and what is known, and has no independent existence and therefore no meaning in isolation. Thus it is nonsense to think of space or time existing beyond the bounds of the material Universe. Time and space are no more than abstractions or metaphors representing the most direct relationships between events and objects. In astronomy, these relationships are expressed in terms of the invariables of the speed and path of light in a vacuum.

In space, relationships are normally symmetrical, whereas in time they are asymmetrical. To talk of time going backward or turning round is a violation of its meaning. Either that, or it is a naive attempt to change the meaning of time from that of an abstract asymmetrical relationship to that of a capricious entity that has a real Platonic existence but needs the material Universe to burst into being before it can kick in, do its baffling stuff, and eventually get reeled in with the collapsing Universe.

Maybe this type of thinking is simply a confusion between events and their relationships to each other. This muddle could account for the perception that time is caused by events and vice versa, rather than it merely representing a set of primitive relationships between events such as concurrence, precedence, and antecedence. Although this may beg the question of how terms like "precede" can have meaning without the concept of time, I do not believe that one can coherently separate generalizations such as time from ideas like "before" or "after" because they are fundamentally the same a priori concept. Thus a phrase like "before time" is complete nonsense, unless one is using it in some metaphorical or poetical sense to express the profound nothingness before the Universe came into existence, or unless it stands for ideas like "too early." Otherwise it is worse than saying "before the before."

This discussion would be completely pointless and even meaningless if in cosmology our ideas of space and time were not defined as a relationship determined by the behavior of light. It would also be pointless if light's behavior were uniform throughout the Universe, or if we had no way of realizing that it varied from place to place. But light does not behave uniformly in a vacuum, and we know this because we can make comparisons and because we can, in extreme cases, actually see light being distorted by dense masses. Since the path light travels is defined as the shape of space itself, then in this limited sense it is meaningful to talk of time operating at different rates relative to an outside observer, and of space being curved. Also, as I will show, this concept of distorted space has significant practical consequences in our investigations of the nature and extent of the material Universe. It is therefore, as it happens, the best means by which to apprehend the overall shape and destiny of our Universe.

Einstein's vision not only provides us with a coherent view of the Universe, but also gives observational astronomers direct practical means with which to investigate his idea of reality. However outlandish and incredible Einstein's theories may seem, they are straightforward when interpreted and used in the way I think he intended. Because we enjoy imponderables and playing around with metaphysical ideas, there is always the temptation to confound ourselves with counterintuitive interpretations of what are often in practice remarkably sensible and straightforward insights. Like all truly great and far-reaching ideas, Einstein's conceptual breakthrough is in essence very simple. It is exceptionally useful to those who prefer to get their hands dirty with actual exploration and measurement of external reality, rather than grandiloquent theorizing or dwelling on untestable cosmologies. He provided us with a different way of regarding the Universe by giving us a comprehensive and coherent metaphor for space and time. This metaphor is simply an expression of a better way of doing our sums, so that we could do away with many problems and paradoxes that we were running up against in trying to measure the Universe. As far as observational astronomers are concerned, the core of Einstein's insight was that we were having problems because we failed to appreciate that light is influenced by mass and therefore cannot travel in the intellectual abstraction of a straight line. Apart from anything else, this gave astronomers a worthwhile objective: to find observable instances of light being distorted by the mass of compact bodies. Initially the task of detecting such gravitational lensing events was undertaken in order to confirm Einstein's theory. However, it came to be appreciated as the most effective means for detecting the presence of dark objects such as black holes.

If, as many astronomers believe, a significant proportion of dark matter is in the form of compact bodies, then both the problem of the missing mass and its possible solution would be a gravitational issue and therefore a concern of observational astronomy. Not only did the problem arise because the presence of dark matter became apparent through its pervasive gravitational influence, but a solution to the identity of the missing mass might lie in the fact that an individual dark body can betray its presence by its gravitational effect. Einstein's insight that dense masses distort light is therefore crucial to providing astronomers with a means for identifying dark matter. In the event, such "gravitational lensing" turns out to be far more important for finding the missing mass of the Universe than even the most optimistic astronomers anticipated.

10

DARK GLASSES

Although the narrow corridor appears
So short, the journey took me twenty years.
     — THOM GUNN, *"The Nature of an Action"*

GRAVITATIONAL LENSING is the ultimate way of finding dark matter. This practical application of the theory of general relativity is related to the effect that Eddington sought to detect in his solar eclipse observations when trying to confirm Einstein's theory. The gravity of a sufficiently massive and dense body will bend the light of a more distant luminous object along the line of sight. Obviously, for this effect to be discernible the massive body itself cannot be too bright. So Eddington had to wait for the Sun to be darkened during a total eclipse so that he could observe the lensing of a conveniently placed star. He claimed that this star showed the expected displacement due to the gravitational bending of its light rays by the Sun.

A more dramatic manifestation of gravitational lensing is seen when a massive compact galaxy lies along the line of sight to a quasar. In this case the galaxy behaves rather like a type of glass lens bending light with its gravitational field, so that the quasar splits into two or more separate images. The multiple images of a famous system known as the Einstein Cross is a beautiful example of such "macrolensing." Although the lensing galaxy is not always easily visible, we can be sure that it is there somewhere. When we do observe it there is much that we can learn about its mass distribution. Even more spectacular is when a massive compact cluster of galaxies lenses the galaxy population beyond. The enormous gravitational field has the effect of distorting the galaxies behind the cluster. In favorable circumstances these galaxy images are spun out into long arcs. Again, analysis of the distortions can be used to reconstruct the mass distribution of the cluster. A similar effect can be achieved by an invisible lens such as a massive black hole. Thus an otherwise undetectable entity can betray its dark presence by its distorting gravitational effect on the light of a bright object.

Smaller masses can also gravitationally lens the light of more remote objects, but the discernible effect is different. The smaller scale of the lens, combined with random differences in velocity, means that the configuration of source and lens relative to the observer can change over months or years. Thus overall changes in brightness can be observed, even though the detailed distortions of the image are too small to be seen. A distant luminous compact body will therefore appear to change in brightness when lensed by a relatively low-mass object. Such microlensing light curves can result in continuous variations in brightness as lenses crisscross the line of sight to the source. This is in contrast to macrolensing, which presents an essentially static picture because it is on a much larger scale and therefore seems unchanging. A macrolensing event could persist for millions of years, whereas a microlensing event could last for as little as an hour.

The problem with observing microlensing events is how to distinguish intrinsic variations from those caused by gravitational lensing. When we look at a variable star for instance, it is not always easy to tell whether it has reasons of its own for changing in brightness, or whether it only seems to be varying because its image is magnified by the gravity of an intervening dark mass. Provided one has a long enough run of observations, variable stars can usually be eliminated due to their well-defined periodic variations, although some vary irregularly or only occasionally. The latter include red and blue supergiants, novae, and other cataclysmic variables.

The phenomenon of microlensing has up till now only been considered useful by astronomers looking for relatively rare microlensing events within the halo of our Galaxy. These lensing bodies are usually assumed to be low-mass stars or brown dwarfs. Hence microlensing could be an important means for detecting the small atomic proportion of the missing mass. Indeed, at present it is the only plausible way of finding profoundly dark bodies. Astronomers have now witnessed a number of these rare microlensing events, in which a star in the outer parts of our Galaxy lenses a more distant star in a nearby dwarf galaxy.

Microlensing events are always difficult to recognize for certain. There remains the possibility, and in many cases the firmly held belief, that any variation in brightness is intrinsic to the object being observed. A good example of this uncertainty concerns quasars, which become enormously bright by accelerating a whirlpool of matter to close to the speed of light, which then plunges through the event horizon of a very massive black hole in their hearts to effectively disappear from our Universe. From soon after the time of their discovery, quasars have been known to vary in brightness, and it is generally accepted that this variability is intrinsic to the quasar itself. For example, variations might be caused by ripples in the whirlpool of matter in the accretion disk, or perhaps the occasional star lighting up as it accelerates through the event horizon of the black hole at the center.

Astronomers can be mesmerized by such acceptable though unsubstantiated beliefs. For example, I had been puzzling over quasar variability for about 16 years without even considering the possibility that their behavior might not be intrinsic. I started to question this long-held prejudice only when a friend Rachel Webster suggested that the expected profile of microlensing events might match my quasar light curves. It then occurred to me that, just possibly, we had all along unwittingly been observing ubiquitous microlensing events. At first sight there did not seem to be any way of testing this startling idea, apart from pointing out apparent similarities between observed light curves and numerical simulations of microlensing. However, microlensing is a precise geometrical phenomenon and one can make a number of very specific predictions about the properties of the resulting light curves. For example, the way objects change color as they vary will typically be very different depending upon whether the variation is intrinsic or caused by microlensing. The predicted properties of microlensing light curves match very well with those present in what we actually observe.

The implications of the idea that nearly all quasars are microlensed are far-reaching. A huge proportion of dark matter must occur as compact bodies which, because of the constraints imposed by the theory of nucleosynthesis, must be nonatomic entities. In addition as we have seen, there is a well-established theory put forward by Bill Press and Jim Gunn[1] which states that if the density of the Universe is close to the critical density, then every line of sight will be microlensed. This accords with the theory of inflation, which requires such a flat Universe. Whichever way one looks out into the Universe, if there happens to be a quasar along the same line of sight, it will be variable. If these variations are microlensing events, then the ideas of inflation are validated and we know the geometry of the Universe as well as the extent and identity of its material content. In other words, the discovery that most quasars are being microlensed would confirm the widely held belief that there is 100 times more matter than has been observed. The implications are, however, completely unexpected to the extent that they falsify the equally widely held belief that the main constituents of dark matter are elementary particles. Not only does quasar microlensing inform us of roughly the amount of dark matter, but it also tells us that this dark matter is composed almost entirely of compact bodies.

My evidence that dark matter occurs in compact form conflicts with the idea of elementary particles as dark matter candidates. These particles are not expected to form into discrete compact bodies. There is a strongly held prejudice that what is compact can only be atomic in origin. Because of this assumption a great deal of effort has been invested in trying to find ways of either reducing or removing the missing mass problem. For example, there have been many brave challenges to the remarkably robust theories of inflation and nucleosynthesis. The recent large values obtained for the Hubble constant have not helped this endeavor, since increasing the value of this parameter lowers the amount of atomic material that can be formed. Also, a large value has the effect of decreasing the amount of matter predicted from studies of galaxy kinematics.

So, leaving aside for the moment such challenges to our generally accepted ideas, if quasars are being microlensed then all other dark matter candidates are effectively ruled out as significant contributors to this hidden mass. The combined weight of ordinary atomic dark bodies such as brown dwarfs would fade into insignificance since they would add little to the visible stuff. Similarly, the relevance of the search for exotic elementary particles would be diminished. Although the possibility of MACHOs and WIMPs would be interesting to us, rather in the way that the planets are considered important, they would be almost irrelevant in the context of weighing the Universe. They could contribute very little to theories about the large-scale structure of the Universe, galaxy formation, and what happened in the big bang.

Like all comprehensive ideas, this whole business of the problem of dark matter and my solution to it has an element of circularity. Although historically the dark matter problem started with astronomers discovering through the study of star and galaxy kinematics that a great deal of matter is hidden, it could just as easily have arisen if we happened to notice that every line of sight is being microlensed. What I now consider to be the solution could have been just the start of the missing mass problem. Although the problem of dark matter would have assumed a different character, all we would have needed for it to arise is the theory of microlensing, and the currently undisputed proposition that if every line of sight is microlensed, then the Universe is close to the critical density.

As things stand, the validity of my solution crucially depends on the validity of these theories. My conclusion that nearly all quasars are being microlensed supports the theory of inflation, and is also consistent with the most recent mass estimates from analyses of the dynamics of galaxies. The theory of nucleosynthesis and all the related ideas about what was cooked up in the furnace of the big bang help describe the nature of the compact bodies causing the microlensing, but apart from lending some credence these theories of the early Universe have little bearing on the core of my idea. My main result is simply that the density of the Universe is close to the critical value, and most of its material is in the form of dark compact bodies of the right mass and in sufficient quantities to cause microlensing of quasars.

What can we say about the mass of these compact bodies? Microlensing theory provides the means for making more precise estimates of their mass. We can measure the Einstein radius. This is the distance from the line of sight within which the gravitational field has a significant bending effect. It tells us how massive the lensing object is, but it tells us little about the object's density or compactness. Thus, in the bare factual report of my discovery published in *Nature* in 1993,[2] all I could say about these ubiquitous microlensing bodies is that they are dark, Jupiter-mass objects. Any more would have been a matter of interpretation and argument, which I felt was inappropriate for an initial report in a scientific journal like *Nature,* which favors brief and strongly supported ideas with no hint of what the editors regard as speculation. (All the same, papers in *Nature* are expected to express new and even provocative ideas. The requirements for publication are therefore extremely demanding.)

However, when some close colleagues and I publicized my findings in the mass media, we were able to be a lot more expansive and controversial. The unadorned and sparse conclusion was that 95 percent of all matter comes in the form of dark bodies that are about the same mass as the planet Jupiter. Obviously, it is possible to challenge the validity of the research and the methodology that led me to this result. I could be making an error when deciding that quasar variability is the manifestation of microlensing events. However, I do not believe that this technical aspect of my work could properly be described as conjectural. The conclusion might be wrong or surprising, but it cannot by any stretch of the imagination be described as speculative. Its implications for the dark matter problem and the material content of the Universe are hard to avoid. Consequently, though a diminishing number of astronomers simply refuse to believe that I could have produced such an unexpected result legitimately, an important area of debate lies in deciding just what these Jupiter-mass bodies could be.

Some of my colleagues can find arguments for challenging the received wisdom, and maintain that regardless of how much matter is missing it could nevertheless be in the form of atomic material. So these Jupiter-mass bodies could be rather like ordinary planets. But this would pose a formidable problem in working out how such a population of objects could have formed. This does not mean that the idea is wrong: arguments from personal incredulity are weak, and almost always end up serving as an embarrassment to those who try to use them. Nevertheless, there is no observational evidence and very little theoretical support for the proposition that these microlensing bodies could be uniformly sized, planetary objects. However, if in accordance with standard nucleosynthesis theory these objects are taken to be nonatomic, then the only thing they could be is primordial black holes. Their existence, their abundance, and their uniformity are all consistent with generally accepted theories of what happened in the early Universe. Not only is there a plausible mechanism for forming primordial Jupiter-mass black holes, but this idea has significant potential for enriching and modifying theories about the early Universe.

So, despite the outrage of one or two of our astrophysical mandarins who feel that it is beyond the domain of observational astronomers to engage in such speculation, we went ahead and publicized the convincing and alluring idea of primordial black holes as the main constituents of the Universe. The reaction of one of our most prominent astrophysicists was terse. He said, "Rubbish!" Another highly respected and influential mandarin publicly expressed his conviction that quasar variation must be intrinsic, and that my idea therefore had to be the result of bad research. The rest of the astrophysical community was more circumspect, and sensibly chose to reserve judgment. They resisted the temptation to rely on the weight of their authority to lend validity to their personal incredulity. For example, Stephen Hawking and Princeton University's Bohdan Paczynski acknowledged that it was plausible, and could be very important if it turned out to be correct.

At the time we were alarmed by the display of disapproval by one of our leading astronomers. In consequence, I lost the support of a close colleague who until then had shown the greatest enthusiasm and imagination in developing and publicizing the idea. His reaction to the attack was to issue a public proclamation expressing reservations about the research and our interpretation of the results. He justified this on the grounds of the apparently spurious statistic that it had a chance of only 1 in 10 of being right. Sensing blood, the Sunday newspapers had a field day hyping up what in reality had very little to do with science, and a lot to do with astronomers naively trying to hold on to the ephemeral status of being administrators.

The majority of interested astronomers were objective and professional when expressing either support for, or skepticism toward, the idea that the vast bulk of the Universe consists of compact, Jupiter-mass bodies. Both supporters and skeptics contributed a great deal to the idea. Even those determined to prove it wrong ended up making it more robust. I am especially impressed by the exhilarating attitude of the German astronomical community. They were the most rigorous critics and the most uncompromising supporters of the validity of my idea.

Why are primordial black holes the only candidates for the myriad of microlensing bodies? These bodies could be something that we have not yet thought up, but primordial black holes are at present the only speculative entities that can plausibly exist in such abundance at a uniform mass. According to our ideas of the early Universe, primordial black holes can form when the Universe condenses from one state to another as it cools, rather like steam condensing into water droplets. When this happens, the resulting turbulence can result in matter collapsing into black holes. Material within the event horizon of a suitable perturbation will disappear into a singularity. The size of the event horizon depends on the age of the Universe at that moment and this process produces a population of black holes with a fairly uniform mass related to the typical amount of matter within each event horizon. The point at which elementary particles known as quarks condense into atomic type material (hadrons) is called the quark-hadron transition. This took place within the first millionth of a second of creation, and David Schramm has argued that a volume about the size of a beach ball containing an amount of matter roughly equal to the mass of Jupiter could readily form a black hole. The astonishing implication of this is that the mass of Jupiter, which is itself a thousand times more massive than the Earth, would be compressed to a size almost small enough to get your arms around.

Thus we can describe the creation of primordial black holes in terms of particle physics. Nevertheless, they are primarily cosmological entities whose existence is explained in terms of space and time, and therefore in terms of Einsteinian gravitation. Unlike atoms, they are not held together by nuclear physical processes. Their formation depends on an absolutely unique set of circumstances which came together at a specific, unrepeatable moment in the history of the Universe. This could never happen again, because the entire Universe and therefore the totality of space and time were implicated in their formation. In a way, they are like a multitude of inaccessible fossils whose secretive dark presence represents the physical state of everything at that point in the evolution of the cosmos. Each primordial black hole bears the imprint of the Universe as it was within the first millionth of a second of its creation, when these black holes were formed.

Although they are probably composed of nothing stranger than quarks, primordial black holes nevertheless represent a completely different order of being to the rest of the material Universe. Apart from their gravitational influence, they have always been completely out of touch with everything else, and always will be. Also, apart from changing location they are inherently immutable. The only force or law that they are subject to is that of gravity. In other words, they are in themselves unsusceptible to anything except the shape of space. Nevertheless, having said all that, they could lose their pristine identity in catastrophic circumstances such as collisions with another massive body. It is conceivable that some primordial black holes could have fallen into each other, making more massive black holes, but such collisions would be extremely rare. Typical separation between these primordial black holes would be about 30 light-years, some 10 times the distance to the nearest stars. Unlike stars, however, they would permeate all of space. Their overwhelming presence in the vast dark reaches of the Universe would account for their enormous contribution to the mass of the Universe.

11

COSMIC BEACONS

We shall not cease from exploration
And the end of all our exploring
Will be to arrive where we started
And know the place for the first time.
— T. S. ELIOT, *"Little Gidding,"* Part III

IN THE EARLIER CHAPTERS WE HAVE ASSEMBLED the pieces of a fascinating jigsaw, and in the preceding chapter we have seen an overview of the end result, the "picture on the box." The final task is to fit everything together and look in detail at what has been achieved.

The story began in 1977, shortly after the completion of the exciting new UK Schmidt Telescope, at Siding Spring Observatory in New South Wales, Australia. This is a specialized instrument with a very wide field of view for surveying large areas of the sky on photographic plates. During the 1950s, the largest Schmidt telescope in the world was constructed at Mount Palomar in California. This 1.2 meter celestial camera was used to survey the whole of the northern sky on 35 cm square glass photographic plates. This Palomar Sky Survey was carried out on blue and red sensitive plates and detected much fainter objects than any previous survey. The advantage of having both red and blue passbands is that the colors of stars and galaxies can be measured. Thus red stars would appear brighter on the red sensitive plates than on the blue ones, and vice versa for blue stars. The Palomar Sky Survey formed the basis of a large number of important projects and played a major role in our understanding of the large-scale structure of the Universe. However, it could only cover the northern sky, and there was an ever increasing demand from astronomers for a similar survey in the south.

In the event, two Schmidt telescopes were constructed in the southern hemisphere: the UK Schmidt, which was a near copy of the Palomar instrument, and a slightly smaller 1 meter telescope built by the European Southern Observatory at La Silla in Chile. The ESO Schmidt was actually completed first, and the original idea was that it would survey the sky in a red passband, while the UK Schmidt would survey it in the blue. As it turned out, the ESO Schmidt ran into one problem after another, but although they eventually fulfilled their side of the agreement, it was the UK Schmidt that dominated the field from the outset. There were several reasons for this. For example, for safety purposes it was not permitted at La Silla to use hydrogen to sensitize the photographic plates, which caused tremendous difficulties in achieving the same quality and depth as the UK Schmidt plates. Because of the size and optical design of the telescope, exposures had to be nearly twice as long as on the UK Schmidt, which left that much more time for something to go wrong with any individual plate.

The UK Schmidt was a resounding success from the start. Under the direction of Russell Cannon, high quality standards were laid down for the survey of the southern sky and maintained throughout. At the same time a vigorous program of "nonsurvey" projects was undertaken, and astronomers were encouraged to submit proposals for work outside the main survey. This usually meant a more detailed study of a small area of sky, typically to be photographed in a number of passbands. The telescope had been built with an achromatic corrector plate that enabled photographs to be taken in the near ultraviolet, through blue and red passbands, to the near infrared without distorting images.

Many exciting and innovative proposals were put forward by astronomers from Britain and throughout the world. These included projects to delineate the structure of our Galaxy, huge galaxy surveys for mapping out the large-scale structure of the Universe, and searches for distant quasars to try to understand the evolution of these enigmatic objects with cosmic time. By contrast, one of my programs was quite modest in its aims. My idea was to concentrate on one particular Schmidt field, an area of about 6 degrees on a side, and to monitor this field over a period of time to look for variables. I chose a field that was not special in any way, with no dominant galaxies or star clusters. The field also had to be in a part of the sky where competition from other programs was likely to be minimal (I was hoping for lots of plates!), and not too near the galactic plane, where the overcrowding of stars would have made the plates very difficult to analyze. The field that I finally chose was called ESO/SRC 287.

A deep photograph from the UK Schmidt Telescope is an amazingly beautiful artifact. At first glance we see a 35 cm square of glass with a few dark patches on a gray background. Under the microscope a new world is revealed. The background is resolved into a myriad of faint stars and galaxies, black on gray as astronomers normally work with negatives. The larger dark patches are revealed as the images of bright stars, galaxies, and star clusters. One can spend hours exploring a Schmidt plate under a microscope, and will inevitably be assailed by a stream of intriguing questions.

Astronomical photography goes back to the last century, when the first attempts to photograph the sky through a telescope were made. At first the main idea was to use it to obtain relative star positions so that they could later be conveniently measured, but it was obvious right at the outset that this was a way to see fainter stars and galaxies. The eye has an effective integration time of about one-tenth of a second. In other words, this is the time available for an image to build up on the retina before it is "wiped," leaving the retina ready for the next image. Photographs, on the other hand, can be exposed for many minutes or even hours, allowing much more light to be recorded. The measurement of the brightness of objects on photographs has always presented a problem, as the amount of blackening on the emulsion is a complicated function of the amount of light falling on it. Nonetheless, in the first half of this century a number of ingenious methods were developed to circumvent the problem, and photography was the standard way of making astronomical observations.

Eventually better ways were found to detect the light from astronomical objects. At first systems related to television cameras were used and now we use detectors known as charge coupled devices (CCDs). The advantage of these new devices is that they are much more sensitive, and their output is related in a simple linear way to the amount of light that falls on them, minimizing the need for extensive calibration. However, they all suffer from the major drawback that they cover only a small area of sky, and so are quite unsuitable for survey work. Schmidt telescopes, with their huge field of view, therefore still have to use photographic plates as detectors.

Early projects based on Schmidt plates required them to be scanned by eye, for example to count star numbers in various patches of sky, to catalogue galaxy clusters or to search for unusual looking objects. Projects like these were extremely time consuming, sometimes involving thousands of hours of peering down a microscope, and inevitably suffered from the consequences of human frailty in the form of inconsistency and bias. This was the background to one of the great developments of British astronomy since the Second World War. The idea was to build a machine that would automatically scan photographic plates, detecting stars and galaxies above some predetermined threshold, and measure a number of features or parameters describing them. A high priority was to be placed on speed, so that large numbers of plates could be measured in a reasonable amount of time.

The first machine of this type was the inspiration of Vincent Reddish and was built at Edinburgh's Royal Observatory in the 1960s. It was called GALAXY, and was capable of measuring the positions of stars with great accuracy. GALAXY was rather limited in its other measuring capabilities, and was not suitable for a number of important cosmological projects that astronomers were eager to undertake. This prompted Reddish to propose a new machine to be named COSMOS, which would be able to measure not only the positions, but many other parameters describing each image, including brightness, shape, and size.

COSMOS was built at the beginning of the 1970s, and was completed shortly after the UK Schmidt was fully commissioned and had started the Southern Sky Survey. Unfortunately, COSMOS was bedeviled by teething problems, and it was several years before the quality of the data it produced was acceptable. Whether or not this was due to cutting corners in development in order to be ready for the commissioning of the UK Schmidt, the consequences were quite serious. Astronomers learned to distrust the early data, and it was some time before there was a general acceptance that the measures could be used with confidence. This is a problem that occurs again and again with new instruments. Should one release them for general use in an attempt to meet deadlines before ensuring that everything is working properly, or should one try to iron out every little bug at the expense of running over time and failing to meet commitments? To make matters worse, a similar machine was built at Cambridge University called the Automated Plate Measuring Machine (APM), and there has been an uneasy rivalry between the two machines ever since.

My plan to monitor the variables in field 287 was crucially dependent on measurements from COSMOS. I first of all chose a high-quality plate as "master," which when measured by COSMOS produced a catalogue of about 200,000 stars and galaxies. The COSMOS brightness measures were converted to the standard system of magnitudes using sequences of stars from CCD observations, and then all subsequent plates were measured in a similar way and paired up with the master. Thus, for each of the 200,000 objects a string of magnitude measures was obtained covering various time intervals. Variable stars were then just those which changed significantly in magnitude.

To start with, it was not quite as easy as this. Various types of measurement error acted to mimic the changes in magnitude of the variables, producing a confusing background of spurious candidates. Eventually the various sources of error were sorted out. For example, it turned out that many of the so-called variables were galaxies and not stars at all. These galaxies were not true variables, but because of their diffuse images the accuracy of the photometry from plate to plate was not very good, and they showed up as variable. This problem was solved by devising a reliable method of eliminating galaxies on the basis of their more extended images. After all, I was not looking for variable galaxies, but stars.

The very first project I undertook with the data was a bit of a long shot, and although the results were somewhat inconclusive I still think it was a good idea and would like to pursue it again in the future. It concerned the search for very distant supernovae. These exploding stars, apart from being interesting in their own right, have always been important to cosmologists. For a few days the huge energy of the explosion causes them to outshine an entire galaxy, and enables the event to be seen at a great distance, thus making it an ideal cosmological tool. The difficulty has always been to find high-redshift supernovae, and a number of groups have devised methods for early detection of such events that could then be monitored on large telescopes. My approach was rather different. I planned to take a photograph of field 287 on every possible night over a period of two or three weeks. I calculated that, among the 100,000 or so galaxies in the field, at least one should contain a supernova explosion that could be followed during the period of observation. Because of the difficulty of finding a starlike brightening in a distant galaxy from the rather sparse information available in the COSMOS measures, this was the first and last time I carried out a search of the field by eye. It certainly brought home to me the tremendous benefits of an automatic measuring machine, and made me full of admiration for the early pioneers of Schmidt plate astronomy, who had achieved so much with their eyes alone. After several weeks I found one candidate that seemed to match the profile of a supernova explosion really well. What I was interested in was the decay time, the rate at which the supernova became fainter. Nearby supernovae are known to halve their brightness every six days, the time after they have passed their maximum luminosity. However, a supernova at high redshift should appear to get fainter more slowly due to an effect known as time dilation. This is one of the consequences of general relativity, and bears some similarity to the Doppler effect, which we have already met. The idea is that when we observe an object at high redshift, we see time running more slowly than on Earth. If we observed a clock in a high-redshift galaxy and compared it with an identical one on Earth, it would appear to be running more slowly. The situation is symmetrical in the sense that an observer in the distant galaxy would see our clock running slower than his. At first sight this might seem like a logical contradiction; in fact it is not, but is typical of the unfamiliar situations that arise in the realm of general relativity.

My plan was to see if I could observe this time dilation effect, and in fact my high-redshift supernova did decline in brightness more slowly than expected. However, one cannot generalize from a sample of one, and to get a definitive result I would have needed a number of other examples. Another point that deterred me from pursuing this project further was the fact that all plausible models of an expanding Universe predict the same time dilation factor of $1 + z$, where z stands for the redshift. It is really only some models of static Universes which would not show time dilation and these tend to be beset with other difficulties. There seemed little point in devoting myself to the task of ruling out something that almost nobody believed in anyway, and with good reason. I therefore decided to move on, and follow up some interesting lines of inquiry that were suggested by the COSMOS measurements.

I think the most important part of an observational astronomer's work is to decide what problems to tackle. The aim should be to choose a project that, if successful, will make a big impact on our understanding of the Universe, but which is nevertheless feasible. There is no point in doing work that, however well it goes, will be of little interest to anybody, although such undertakings very easily pass muster in the peer review process and are a good way of clocking up sheer numbers of publications. It is sometimes difficult to credit the level of triviality of some projects that are published in the most reputable journals. But equally, it is foolhardy to embark on a project that in principle could rewrite all the textbooks, but for which the chances of a successful outcome are negligible. For example, I became involved in a project to detect cosmic strings, the speculative entities that may have resulted from a change of state, or phase transition, in the early Universe. These stringlike fractures are essentially breaks in space-time which whip through the Universe at the speed of light, acting as threadlike gravitational lenses splitting in two the images of more distant galaxies. The plan was to look for strings of double galaxies, like beads on a necklace, which would reveal the presence of a cosmic string. Even with the help of the superb UK Schmidt and COSMOS data, this mission was fraught with horrendous difficulties. It would, for example, be almost impossible to distinguish the split galaxy images from normal double galaxies. Although a successful outcome would have been a major scientific breakthrough, I think the project was fundamentally misguided, and it certainly failed to find any cosmic strings. Genuine insight, good sense, and a real feel for the subject are needed to spot a new and important line of research. It also helps if one is not too indoctrinated by the current received wisdom. Once a really good project has been thought up, the rest is usually quite straightforward requiring dogged determination, care, and often a fair amount of political skill to obtain the necessary resources. Admittedly there are a few outstanding exceptions to this, where extremely important results have come from the most unpromising starting points, but I do not think it makes sense to rely on such miracles when deciding where to direct one's energies.

Once the problems with the COSMOS data had been sorted out, a quick search revealed a few stars varying by two or three magnitudes over a few days. These turned out to be objects known as cataclysmic variables, a generic classification for a disparate collection of stars in various states of catastrophic upheaval. One of these stars in particular turned out to be very interesting, but they need not concern us further here.[1] The first systematic survey I carried out with the COSMOS data was a search for RR Lyrae variables. These famous stars have been used in many cosmological projects, most notably the measurement of the Hubble constant, their special characteristic being that they vary in a way that makes them particularly easy to recognize. They have a well-established luminosity, which means that it is fairly straightforward to calculate their distance, so they make ideal markers for mapping out our Galaxy. My plan was to measure the way the star numbers in the halo of the galaxy fall off at great distances, using RR Lyrae variables as tracers. This involved surveying the whole of my Schmidt field for stars that varied in the characteristic way of RR Lyraes, and counting the numbers at different distances from the galactic center. Because of the great depth of the plates from the UK Schmidt, I was able to find these stars at much greater distances than ever before.[2]

The second part of the program was to measure the velocities of my sample of RR Lyraes. These velocities would be caused by the gravitational pull of the Galaxy. This made it possible to estimate the mass in the halo that was causing the stars to move and was my first attempt at measuring dark matter. The main difficulty with this project turned out to be in overcoming the large uncertainty in the velocities due to the fact that RR Lyraes vary by the pulsation of their outer layers. This meant that there was an unknown contribution to the velocity of the stars from the expansion or contraction of their atmospheres. In spite of this difficulty, the results turned out to be exciting. One of the most distant stars proved to have an extremely high velocity, implying a very large mass of dark matter in the outer parts of the halo, a conclusion that was supported by the average velocity of the rest of the sample. One of the expected features of the distribution of dark matter in the halo is that it should be much less concentrated toward the center of the galaxy than visible matter.[3] This is why the very deep plates from the UK Schmidt were so important. They made it possible to extend the survey to a distance where dark matter should dominate over normal stars, and it was thus possible to show that there was indeed a huge excess of unseen mass in the outer parts of our Galaxy.

As I continued my search, it gradually became clear that short-term variables such as RR Lyraes and some types of cataclysmic variable formed a relatively small fraction of my candidates. This was in contrast to previous surveys for brighter variables where short periods dominate. At that time the plates available in field 287 spanned the two years 1977 and 1978, and I could not help noticing that most of the variables I had detected showed no measurable change over weeks or months, but only varied between the 1977 and 1978 epochs. Furthermore, these longer period variables tended to be fainter than the RR Lyraes, and were more numerous at fainter magnitudes, bunching up toward the magnitude limits of the plates.

I was now confronted by the perennial agony of observational astronomers. One comes across an intriguing object that requires detailed follow-up on a large telescope to understand what it is but first one must obtain an allocation of precious telescope time. This is a highly competitive process: typically only one in five proposals are allocated time because the number of applications is so large. Even if one is successful, there is still a long wait for the few nights of the allocation to arrive. If one is lucky enough to already have some observing time in the pipeline, it is often possible to use a small part of this to follow up new ideas. Although this is officially frowned upon, it is generally accepted that in moderation it is a useful way of kick-starting promising new projects.

In the case of my variables, I had to wait some 18 months before I arrived at the 3.9 meter Anglo-Australian Telescope at Siding Spring, on the same site as the UK Schmidt. The AAT was one of the largest telescopes in the world, and well suited to obtaining spectroscopic observations of the variables to determine what they were. This involves spreading the light from an object into its different colors, as with a prism, and looking for light or dark features known as "lines," which can be identified with the light from atoms or molecules making up the surface of the object. The resulting patterns of lines should then be sufficient for a positive identification.

When I looked at the spectra of the first variables, I saw the characteristic broad, bright emission lines of quasars. The lines originate in the outer cloud regions of quasars, typically from hydrogen, carbon, and magnesium atoms, which are made to fluoresce by the intense blast of radiation from the central accretion disk. The lines are very broad because the emitting atoms are moving at large random velocities, and the consequent Doppler shifts smear out the color of the lines. I was intrigued when I realized that most of the variables were quasars. Apart from anything else, variability seemed to be a new way of detecting quasars, and because they have always been such enigmatic objects I felt quite sure that some interesting developments lay ahead.

Because quasars are so luminous, they can be observed at very great distances. We have already seen that the finite speed of light means that when we observe any astronomical body we are seeing it as it was some time in the past when the light was emitted. In the case of the nearest stars light takes a few years to traverse the distance to the Earth, and we are seeing them as they were a few years ago. The nearest galaxies are much farther away, and we see them as they were a few million years ago. Quasars are much farther still, and we see them as they were a few billions years ago. The light from the most distant quasar started its journey to us long before the Earth came into existence, when the Universe was only about 10 percent of its present age and a very different place. Thus the study of quasars provides a way of looking back in time and exploring a much younger Universe. Quasars serve as markers to delineate the Universe, rather as RR Lyrae stars can be used to map the halo of our galaxy. This has meant that from their first discovery a great deal of effort has gone into assembling unbiased samples of quasars to map out the Universe, both in space and in time.

The earliest surveys relied on the fact that some quasars are radio sources, but this is rather an inefficient approach since most quasars are "radio quiet," and so other methods of finding them were developed. One difficulty was that quasar images, as observed for example on Schmidt plates, are completely starlike, giving no hint of their fundamentally different nature. Ways had to be devised to distinguish them from the overwhelming majority of background stars. The most successful of the early methods was to use the property of "ultraviolet excess." This exploited the fact that, compared with most stars, quasars emit an excess of light in the bluest color band observable from the Earth, the so-called U-band, close to the point where the atmosphere cuts off ultraviolet radiation.

It was relatively easy to carry out a survey for ultraviolet excess quasars in field 287. This was done by taking photographs in the U-band and looking for starlike objects that were brighter on these plates than on the blue, or B-band, plates that had been used to detect the variables. The main idea at this stage was to estimate what proportion of quasars had been detected as variables. It turned out that only about 2 percent of the ultraviolet excess quasars had varied between 1977 and 1978, not a very efficient process![4] It seemed likely that more time was needed to allow the quasars to vary over their full range, so the survey was restructured to ensure that several plates were taken every year from then on. The plan was to continue monitoring until the long-term pattern of variation became apparent, and at the same time to find out what percentage of quasars were variables.

It eventually became obvious that, contrary to general belief, quasars varied predominantly over very long timescales, and that after about 10 years nearly all quasars had varied enough to reveal their presence on Schmidt plates. This was a great surprise. One of the original puzzles about quasars was that they varied on a short timescale of a few months. It now appeared that this behavior was confined to a very small set of so-called optically violently variable quasars, where it was believed that we were observing the quasar directly down its polar jet, the large rapid variation being caused by instabilities in the outflowing plasma. Normal quasars also varied on short timescales, but only in the shape of small-amplitude flickering that was completely overwhelmed by the long-term variations.

The need to wait for several years while the quasar variations were monitored gave me the opportunity to look at another aspect of the missing mass problem. It has been known since the work of Jan Oort in the 1930s that star velocities in the disk of our Galaxy imply the presence of more matter than we see in the form of stars, gas, and dust. The deficit is not large, about a factor of 2. The most likely candidates for this missing matter have always been low-mass stars, which are so faint that they escape detection in surveys, or perhaps the failed stars known as brown dwarfs. When a normal star is formed, a collapsing gas cloud fragments into smaller pieces or "protostars." These protostars continue to collapse, and in the process the pressure of the gas rises so that they heat up. This is exactly the same phenomenon one observes when pumping up a bicycle tire: as the air is compressed through the valve, it becomes hot. Providing the protostar is more than about 8 percent as massive as the Sun, it becomes hot enough for thermonuclear reactions to take place, similar to those powering a hydrogen bomb. In this way a star is born. The energy produced by thermonuclear reactions has the effect of halting the gravitational collapse, and the star then remains stable for an extended period (around 10 billion years in the case of the Sun) while it consumes its hydrogen. If the protostar's mass is less than 8 percent of the Sun's, the collapse will not generate sufficient heat to start thermonuclear reactions. The protostar will glow briefly from the heat of the collapse, and then rapidly cool and contract to what is known as a degenerate state, where the atoms are packed tightly together but where the mass is far from sufficient to produce a black hole. These dimly glowing low-mass objects, or brown dwarfs, are, like true stars, formed from chunks of gas clouds but, unlike stars, are too small to switch on the thermonuclear reactions that illuminate the sky.

In searching for brown dwarfs there are, in my opinion, three main questions to answer. What is the nature of these cool degenerate stars? How are they distributed throughout the Galaxy? Do they make a significant contribution to the dark matter in our Galaxy? A number of different strategies have been employed to find brown dwarfs. A well-tried approach is to narrow the search area by looking for low-mass binary companions to more normal stars. Such systems can be discovered in several ways. The traditional approach, which has also been used to search for planets outside our Solar System, is to look for stars whose proper motions show a slight "wobble," due to their being pulled from side to side by a faint companion as it travels around its orbit. A second, more modern method is to use an extremely accurate spectrograph to detect small velocities toward and away from us as the companion pulls the primary back and forth in its orbit. The third and so far most successful method is to look directly in the infrared where a cool companion would be less overwhelmed by the radiation from the hotter primary. Thus as one looks farther into the infrared, one should see a second image appearing close to the primary.

Although some very cool stars have been found as companions, I do not think that this is the right way to go. Such stars are very difficult to study in detail because they lie so close to bright primaries, which has the effect of contaminating spectra and making photometry very tricky. Perhaps more important, the proximity of a bright primary could well affect the evolution of the companion in a number of ways. In addition, the fact that such surveys are limited to binary systems means that one can learn little about the overall distribution of low-mass stars. It even seems possible that such companions might be planets, in the sense that they were formed as part of planetary systems as opposed to the fragmentation of a gas cloud. So the most important questions about brown dwarfs cannot be answered by pursuing the search for low-mass companions. For this reason I have always favored approaches that seek to find isolated low-mass stars.

Surveys for low-mass stars can make use of two important properties of these objects. It is a general rule of stellar evolution that the lower the mass of a star, the cooler and less luminous it will be. This rule is broken only at the beginning and end of a star's life, when complicated changes occur. The coolness of low-mass stars results in their being very red with much brighter images on photographic plates which have a redder passband. Thus they can be detected by searching COSMOS measurements of Schmidt plates for stars which are much brighter in the near infrared or "I-band" than in the red or "R-band." The other property that can be used is low luminosity. If a low-mass star or brown dwarf can be detected at all, it must be nearby, and generally speaking so close that its apparent motion relative to more distant stars (known as proper motion) can be observed. Proper motions are measured on Schmidt plates by photographing the same field at several different epochs covering, say, 10 years. Then it is simply a matter of looking for stars that are moving relative to the background of more distant stars. The two properties of red color and large proper motion taken together provide the unambiguous signature of a low-mass star. In practice, the very red color alone is often sufficient to detect a useful sample.

Stars detected by these methods will typically be single and suitable for intensive photometric and spectroscopic study. More important

153

from my point of view, they can be used to understand the part that low-mass stars play in the makeup of our Galaxy. Traditionally this has been done by constructing what is known as a luminosity function. This is basically a tally of the number of stars of progressively smaller brightness or luminosity and can be obtained from an unbiased survey of stars. The main difficulty is in converting the apparent brightness as measured from a Schmidt plate to the intrinsic brightness, or the absolute magnitude. Sometimes this can be done using parallax to measure the distance, which is all that is needed to turn apparent into absolute magnitude. Otherwise, at the expense of much larger errors, one can use a relation between color and luminosity based on the colors of parallax stars. The result of this exercise for normal stars has been known since the 1930s. Numbers increase steadily toward a maximum as the luminosity decreases. Until recently it was believed that the numbers then plummeted to zero. There appeared to be a cutoff in luminosity.

As I became more familiar with the objects in field 287, I realized that there was a large number of extremely red stars with high proper motion, which seemed to be at variance with the idea of a cutoff. I enlisted the collaboration of Mike Bessel, an Australian friend and an expert on low-mass stars, and we went about making our own measure of the luminosity function. My suspicions proved to be correct. We found that, although the luminosity function did reach a maximum, there was no cutoff, and we detected what appeared to be a new population of low-luminosity stars. At the time we thought they might be brown dwarfs. Even today, their exact nature is not entirely clear.

Actually, what one really wants to measure is not the luminosity function but the mass function. This is based on the same idea, but counts the stars in ranges of progressively smaller mass. The mass of a star is not easy to measure directly. Most of our knowledge of stellar masses comes from binary systems, where a detailed knowledge of the orbit enables one to calculate the masses of each component. Although the whole process is very uncertain, this knowledge can be used to derive a conversion between luminosity and mass. Nonetheless, if one does the best job one can, it appears that the observed luminosity function is transformed into a steadily rising mass function. This leaves room for the possibility that if the mass function continues to rise into the brown dwarf regime, it could account for the missing mass in the galactic disk.

One of the reasons why true brown dwarfs are so difficult to detect is that they shine appreciably for only a short time while they are collapsing. Thereafter they are close to invisible at optical wavelengths, only glowing dully in the infrared. If this makes it difficult to detect them in the galactic disk, it makes it almost impossible in the halo. The stars in the halo were all formed at the birth of the Galaxy, and so any brown dwarfs would long ago have cooled to a nearly undetectable level. The situation is made worse by the fact that the halo is much more diffuse than the disk, so the nearest halo stars are much farther away than the nearest disk stars. The missing mass problem in the halo is much more acute than in the disk. Since low-mass stars have always been attractive candidates for the hidden material of the Galactic halo, it is particularly frustrating that they are all but impossible to see.

This deadlock was broken by an inspired idea of Bohdan Paczynski.[5] Our Milky Way Galaxy is orbited by a number of satellite dwarf galaxies, the biggest of which is known as the Large Magellanic Cloud, or LMC. Paczynski argued that, when we look toward the several million stars in the LMC, occasionally a star in the halo of our Galaxy will cross the line of sight, causing a microlensing event. This is yet another manifestation of gravitational lensing. In this case, as the halo star crosses the line of sight, it distorts the image of the LMC star and magnifies it in a way that can be precisely calculated. The scale of the distortions is far too small for them to be resolved by a telescope, but the characteristic brightening and fading is easily measurable. Furthermore, it is possible to distinguish microlensing from the changes of a variable star. The bell-shaped light curve of a microlensing event can be matched exactly to the theoretical curve, and will be the same in all color bands. Also, it should only occur once in any long run of observations since, for any individual star, microlensing is an extremely rare event.

Paczynski's idea was put into practice by a consortium of American and Australian astronomers, who used a telescope solely dedicated to this project to monitor several million stars in the LMC, night after night, in blue and red passbands. Their detectors were CCDs, which are much more sensitive than photographs but which still provide a large enough field of view due to the very high star density in the LMC. In 1993, after about a year of observation, the group published a classic and beautiful example of a microlensing event.[6] By this time other groups had started similar monitoring programs both in the LMC and also in the dense cloud of stars near the center of our Galaxy, known as the galactic bulge. Before long, several microlensing events had been detected, and it became clear that something rather unexpected was happening.

One of the limitations of Paczynski's idea is that the underlying nature of any particular microlensing event is essentially ambiguous. Several factors determine the length of an event. The most important is the mass of the lens. The more massive it is, the more extensive is the effect of its gravitational field, the so-called Einstein radius. This means that the lens will take longer to cross the line of sight, and thus the observed duration of the event will be greater. But this timescale also depends on unknown factors such as the velocity of the lens across the line of sight, and how close it is to the source. Although we can never be sure about the mass of the lens in any individual microlensing event, it is possible to calculate the average mass and distribution of the lenses by assuming a characteristic spread in velocity, and carrying out a statistical analysis of a number of events. When this approach was applied to the first small sample of microlensing light curves, the conclusion was that the number of events in the galactic halo was barely enough to account for the known population of stars, and certainly did not encourage the idea that the halo was composed of brown dwarfs. On the other hand, to everyone's surprise, the surveys toward the galactic bulge detected far more microlensing events than expected. Putting the two sets of observations together, the most likely interpretation seemed to be that the galaxy was not the shape that we believed. However, the picture dramatically changed when the second year's observations became available. Two of the three original events were discarded, one because it no longer seemed like a microlensing event, and the other because it repeated. The probability of any individual star being lensed more than once in a year is vanishingly small, so it had to be excluded as a microlensing event. On the other hand, seven new events were found, some from reanalyzing earlier observations. The current view in the MACHO camp is that 50 percent of the galactic halo is in the form of bodies with half the Sun's mass, but the difficulty of providing a plausible explanation for this, coupled with the inconsistency with past predictions, leaves most astronomers in the wait-and-see mode.

The star that was responsible for what appeared to be a repeating microlensing event has been christened a "bumper." If this class of star, which appears to mimic microlensing events, is at all common, it would cause havoc with microlensing statistics. My own view at present is that most of the events are probably caused by stars in the foreground of the LMC microlensing other, more distant LMC stars. This is an idea of Kailash Sahu,[7] which even the MACHO team concede is likely to happen. The debate is over the number of events that could be produced in this way. This in turn depends on the exact shape and depth of the LMC, which are only poorly known.

The claim that a large fraction of the galactic halo is in the form of bodies with half the Sun's mass presents formidable difficulties. The most obvious possibility is that they are white dwarfs, the end products of massive stars that have had time to complete their respective life cycles. The problem with this idea is that in the process they would have produced extensive contamination of the galactic halo with heavy elements, which we do not see. Also, where are the associated, less massive stars in such a population which have not yet had time to become white dwarfs? To circumvent this problem, some astronomers in the MACHO team have suggested that the MACHOs may be half solar mass primordial black holes.

I think this is a very good example of the uncertainty that is so often inherent in an exciting new project, and of the swings of opinion from which the public is so often shielded.

12

SEEING THE UNSEEN

Mr. Turnbull had predicted evil consequences, . . . and was now
doing the best in his power to bring about the verification of his
own prophecies.

—ANTHONY TROLLOPE, *Phineas Finn*

S THE SCHMIDT PLATES OF FIELD 287 steadily accumulated, the nature of the quasar light curves gradually became
apparent. They typically showed semiregular undulations, with
peaks separated by 5 to 10 years. This was very hard to explain on the
basis of the accretion disk model for quasars, where variations should
typically last only a few months, and longer timescales would be
rapidly damped out.[1] I pondered this problem on and off for 16 years,
and discussed it with a number of theoreticians working on the structure of quasars, but the issue did not seem to capture anyone's imagination. The general feeling tended to be that, the problem of short-
term variations having been solved, the long timescales were a relatively minor detail. Even so, the problem continued to nag at me, but I
remained baffled until in 1992 I went to Australia for an observing
trip. While there, I attended a local astronomy conference and got
talking to Rachel Webster, who specialized in work on gravitational
lenses. We turned to the subject of quasar variability, and I described
the results of my monitoring program and the difficulty I was having
in explaining the variations. When I described the light curves, she
asked whether I had thought of microlensing. She went on to explain
that there had been a number of investigations, especially in Germany,
into the effect of looking at a distant compact light source such as a
quasar through a cloud of small microlensing bodies like the stars in a

galaxy. The result is that the light undergoes multiple microlensing events as it threads its way through the stars moving around in the galaxy. The German group had modeled this situation on a computer and produced simulated light curves which Rachel remembered as being very similar to my observations.

As soon as I got back to Scotland, I chased down copies of the papers showing the simulated light curves, and to my delight many of them bore a remarkable resemblance to the observations.[2] For the first time I had found a plausible explanation for the quasar variations I had been observing for the previous 15 years. The idea was that the space between the Earth and the quasar would be filled with small bodies crisscrossing the line of sight and microlensing the light from the quasar. These lenses could be in galaxies, but not necessarily so. They might fill the huge voids of space between the great agglomerations of galaxies. Both theory and the computer simulations showed that when several bodies microlens a light source such as a quasar at the same time, they combine in a complicated way to produce large amplifications like those in the observed light curves. Although the similarity was certainly suggestive, it was clearly not enough to make a convincing case for the microlensing phenomenon. Computer simulations can only determine the feasibility of an idea, but can very rarely confirm it. Specific tests were needed.

One rather obvious approach concerns the concept of time dilation. As we have seen in the previous chapter, when we observe a clock or any time varying process at high redshift, it appears to run more slowly. This will apply to quasar light curves if the variation is intrinsic to the quasar. The higher the redshift, the longer the timescale of variation should be. This was an effect that had been looked for in a number of earlier surveys, but had never been seen with certainty. However, the small number of quasars and short time coverage of all previous surveys meant that little significance could be attached to any of the results. At this time, the field 287 sample contained over 300 quasars monitored for 17 years, and was thus far superior to any other survey. The most difficult aspect of the test was to find a reliable quantitative way of measuring the timescale of variation. Eventually I decided to use an "autocorrelation function." This is a way of measuring systematic changes in magnitude over different timescales. When I applied it to the light curves in field 287, it seemed to show conclusively that there was no increase in timescale with redshift.[3] On the face of it, this ruled out any form of variation intrinsic to the quasars but was consistent with the microlensing hypothesis. This is because the variations are caused by intervening objects that are much closer, and hence have lower redshifts. Any time dilation will be at this lower redshift. Because of the curved geometry of space, the lenses will all tend to be at the same redshift, regardless of the redshift of the quasar. So little change in any time dilation effect will be observed as quasar redshifts increase.

The success of the time dilation test was very encouraging for the microlensing hypothesis, and prompted me to think through the implications for the nature of the lenses. It was immediately clear that their masses could be measured from the timescale of variation. The distance to which light is significantly affected by the gravitational field of a microlens is known as the Einstein radius. This length depends on the mass of the lens; it is in fact proportional to the square root of the mass. So, if we can measure the Einstein radius, we can calculate the lens mass.[4] To do this, we must assume an average velocity for the lenses across the line of sight to the quasar. This velocity is usually taken to be 600 km/s, based on observing typical galaxy and star motions. The autocorrelation function gives a sort of average timescale for a microlensing event, and is thus a measure of the time taken for a typical Einstein radius to cross the line of sight. So we have a velocity and a time from which we can calculate a length, the Einstein radius. As we have already seen, this in turn gives us the typical mass of the lenses. When I went through this procedure, I concluded that the lens masses must lie in the range from one-tenth to one-thousandth the mass of the Sun; that is, from slightly more than the mass of a brown dwarf down to the mass of the planet Jupiter.

The second conclusion that can be drawn about the microlenses concerns the amount of matter they represent. More specifically, how much mass do the lenses collectively contribute to a standard volume of space? I discovered that this mass density can be estimated from the elegant result obtained by Bill Press and Jim Gunn dating back to the 1970s, well before the theory of inflation was conceived. They showed that if every line of sight is microlensed, then the combined mass density of lenses must be roughly equal to the cosmological critical density.[5] As we have already seen, this is the density of matter in a

Universe on the borderline between being open and closed, and so is just sufficient to halt the expansion of the big bang without causing the Universe to collapse. This is also the density of matter predicted by the theory of inflation. If the microlensing idea is correct, then, since every quasar is being microlensed, and quasars are distributed at random over the sky, one can affirm that every line of sight will be microlensed. Press and Gunn's result then implies that the combined mass of microlenses must be close to the critical density. This very important consequence of the microlensing theory means that the missing matter can be accounted for in the form of small compact bodies, which would thus make up most of the material Universe.

Needless to say, I was extremely excited by this line of reasoning and decided to write it up as a letter to *Nature,* more as an ideas paper than as a tightly argued result. When I finally received the report from the referees I realized that the idea had been taken very seriously. One referee in particular, Joachim Wambsganss, wrote an extremely constructive assessment. He felt that due to the importance of the result if proved correct, it should not be published without more observational support, and he suggested two further tests of the microlensing hypothesis.

The first point he made was that the nearest quasars to the Earth were sufficiently close that the probability of there being a microlens along the line of sight was quite small. One would therefore expect that nearby, or low-redshift, quasars would not show the long-term variability I had attributed to microlensing. The prospect of carrying out this test made me extremely apprehensive, since I had become quite familiar with the quasar light curves by then and had the firm impression that they all showed the same characteristics. So for an anxious period I feared that Joachim had hit on a fatal weakness of the microlensing hypothesis. A more established idea can absorb such blows, but a radically new proposal has to be absolutely impregnable to survive for more than a few months. But, to my relief and surprise, when I reexamined the sample I found that the only two low-redshift members did indeed have light curves completely different from the rest. In place of the long-term undulations characteristic of microlensing, there were a few isolated spikes, which I took to be intrinsic flares of light from the quasar.

Joachim's second test concerned the shape of the light curves. Isolated microlensing events such as those observed in the galactic halo by the MACHO project show a perfectly symmetrical variation. In the case of the multiple microlensing events postulated for quasar variability, the light curves would be symmetrical in a statistical sense, so that there would be no way of telling which way time was running. On average, the rising and falling parts of the light curves should be indistinguishable. This contrasts with most intrinsic forms of variation, which tend to show a sharp rise in brightness followed by a more gradual decline, and thus are not symmetric. When I carried out this test I found that the light curves were symmetrical, as expected for microlensing. This time the result was no surprise, as inspection by eye strongly suggested that there was no asymmetry.

These two blind tests strengthened my confidence in the idea. In its revised form, the paper presented a strong case for quasar microlensing, and it was accepted for publication by *Nature*. It presented a compelling argument that a great deal of dark matter exists in the form of compact bodies, which act as microlenses for the light of quasars. But what were these strange compact bodies that permeated the Universe in such abundance? When and how were they formed? Although I had come up with good observational evidence for the existence of such entities, the all-important question of whether they could be explained in terms of standard big-bang cosmological theory had yet to be addressed.

By the time the paper was on its way to the printers, I had narrowed down my original broad estimate of the mass of each microlensing body to about that of Jupiter. After a more detailed examination of the timescale measures, I felt that I could exclude brown dwarf masses. This was confirmed by a close colleague of Joachim Wambsganss, Peter Schneider, an expert on microlensing simulations. He analyzed the distribution of amplitudes, the frequency of variations of different sizes, in my sample, with the idea of setting limits on the masses of microlenses.[6] He found that he could rule out low-mass stars and brown dwarfs, but his theoretical distribution for Jupiter-mass objects fitted the observations quite well. This approach was completely independent of the timescale argument. The fact that they both gave roughly the same mass strengthened the case for Jupiter-mass lenses.

So were these microlensing bodies really planets like Jupiter? This depended on exactly how much combined mass they represented. For theoretical reasons which we have already discussed, objects similar to planets could contribute no more than about 10 percent to the critical density. Our understanding of the early Universe, combined with measurements of the cosmic abundances of elements such as helium and deuterium, allows us to calculate fairly precisely the density of atomic material created during the era of nucleosynthesis, when atoms were formed. This baryonic matter can have a value of Ω of only about 0.1. Press and Gunn maintained that if the Universe is close to the critical density in the combined mass of compact bodies, then every line of sight will be microlensed, but it is unclear what the lower or upper limit of omega would have to be in order for the same effect to take place. The question is, how big would Ω have to be for compact bodies to cause every line of sight to be microlensed? If nearly all dark matter came in the shape of compact bodies, and Ω was, say, 0.1, then could every line of sight be microlensed?

At the time, I assumed that the microlenses were baryonic, and that there had to be some way in which this relatively small amount of material could microlens every line of sight. I was reluctant to accept the results of computer simulations done by Joachim and Peter which suggested that Ω had to be larger. They came to essentially the same conclusion as Press and Gunn, but from a different direction. Rather than saying that if every line of sight is being microlensed then the combined mass of the lensing bodies is close to the critical density, Joachim and Peter's simulations said that the mass of lensing bodies must be greater than a certain amount for there to be continuous microlensing. The amount was difficult to relate to an actual value of Ω because to simplify their simulations they modeled a quasar being observed through a single galaxy. This artificial state of affairs is not the situation that one would expect in a general cosmological context where the lenses would be spread out along the line of sight. Although, in a qualitative way, it was apparent that according to their model the contribution to Ω from compact bodies would have to be greater than 0.1, it was obviously unsatisfactory, and easy for me to discount. This helped me cling to the idea of a Universe composed largely of objects resembling the planet Jupiter.

My colleague John Peacock had a strong conviction that Joachim and Peter's model, although artificial, nevertheless approximated reality. He was also keen to dislodge me from the position I was adopting by opening up my mind to more exciting possibilities. First it was necessary to relate the density of lenses in Joachim and Peter's single galaxy to the cosmological density. John spent some time computing simulations along these lines. Although not as detailed and sophisticated as that of Joachim and Peter, nevertheless they did relate to the real cosmological context, thus providing a quantitative estimate for the lower limit to the value of Ω for the combined mass of microlenses. To his surprise, I think, he did eventually convince me that this value had to be at least 0.3, which is far more than the baryonic limit of 0.1, and so I would have to face up to the fact that either my microlensing hypothesis was wrong or that at least two-thirds of the Universe is made up of nonbaryonic compact entities, a quite remarkable prospect. If these entities are not planetary objects, then the most obvious conclusion is that they are nonatomic compact bodies. Astronomers have for some time resigned themselves to the fact that the amount of matter in the Universe is more than the permitted amount of atomic material. As we have seen, this conclusion has been reached via the relatively straightforward procedure of measuring the velocities of galaxies on the large scale, which shows that at least 30 percent of the critical density is present. So if they are composed of nonatomic material, what could these Jupiter-mass bodies be? Certainly not planets.

When I asked John what he thought they could be, he said that the only possibility was primordial black holes. I think he saw this rather as a *reductio ad absurdum*. He was clearly very skeptical of the validity of my microlensing hypothesis. Consequently, although it was the outcome of his own determinedly logical turn of mind, he felt under no compulsion to actually believe his black hole interpretation. As far as he was concerned, the idea of a Universe composed mainly of primordial black holes was no more than an academic exercise in the bizarre. However, by this stage I was becoming confident that the microlensing hypothesis was correct, especially since it had survived six months of the most rigorous peer review in my experience. Also, I was persuaded by John's arguments that the lenses amounted to at least 0.3 of the critical density. So, I had no choice. To his consternation, I was completely awestruck by the simple inevitability of his deductions. I really did believe in this picture of a Universe of primordial black holes, and to this day I am still trying to persuade him to believe this radical and far-reaching idea.

As far as I was concerned, John's conclusion was inspired and presented itself to me as a stunning revelation of what the Universe is really like, but was it plausible that such a vast quantity of black holes of just such a mass could have been formed in the early Universe? To help answer this question, I discussed the situation with Andy Taylor, a friend in the astronomy department at Edinburgh University. Andy was much more enthusiastic about the idea than John, and immediately pointed to a mechanism by which the black holes could have formed, first discussed by David Schramm and Michael Crawford about 10 years earlier.

The most obvious way of making black holes in the early Universe is from primordial density fluctuations. These would be ripples in space-time implanted at the beginning of time. As the event horizon grew larger with the expanding Universe and encompassed ripples of larger and larger sizes, matter would collapse about these perturbations to form ever more massive black holes. The problem with this mechanism for accounting for the microlensing bodies was that a wide range of masses would be produced, whereas what the microlensing hypothesis needed was a population of black holes with masses close to that of Jupiter. Andy Taylor looked in a different direction for a solution to the problem.

As we have seen, a feature of our understanding of the early Universe is that as it cools it passes through a succession of phase transitions, somewhat similar to the way in which steam as it cools condenses into water, which eventually freezes into ice. As the Universe passes through a phase transition, it enters a chaotic, turbulent state where, if conditions are right, black holes can be formed in great abundance. A small density fluctuation triggers a catastrophic collapse in which material finds itself in a black hole. The process can be very efficient, and the whole of space can become honeycombed by black holes with only small amounts of matter surviving in the interstices.[7] These primordial black holes would not count as baryonic matter since they would be created before the epoch of nucleosynthesis and would play no part in the building of atoms. The mass of a primordial black hole depends on which phase transition shaped it, and is related to the amount of matter within the event horizon at that point in the history of the Universe. Crawford and Schramm had pointed out that the quark-hadron transition which occurred when the Universe was 1 millionth of a second old would produce black holes of about the mass of Jupiter, providing of course that conditions in the transition were favorable. This was a very exciting development and provided a plausible backdrop to the microlensing picture.

As the publication date for my paper approached, I discussed the idea of a press release with Harvey MacGillivray, the head of the COSMOS unit. We felt that the paper provided an interesting example of what could be achieved with the COSMOS plate measuring machine and the UK Schmidt Telescope, and was worth publicizing. The press officer was enthusiastic about the idea and asked me to draft a suitable text. I summarized the *Nature* paper in what I hoped was an accessible way. After a couple of iterations, the director asked John Peacock to look it over. Ironically, although John did not believe that the Universe is permeated with primordial black holes, he nevertheless felt that the press release was pretty vacuous without a discussion of its implications, and strongly argued that we include it. On the other hand, although inclined to believe in the black hole idea, I was very reluctant to discuss it since it was not part of the *Nature* paper. Also, although to my mind an inevitable outcome of the quasar microlensing hypothesis, the primordial black hole idea was undoubtedly extraordinary and would seem farfetched to those who were not as familiar as we were with all the arguments. It was clear to us that the ubiquitous compact bodies had to be primordial black holes since the alternative, that they were like ordinary planets, was completely untenable. However, I knew that the rest of the astronomical community would be shocked by a media announcement to this effect. Even so, John persuaded me that since this was so obviously the case, we would do more harm by not mentioning it. We did not want to appear incapable of seeing what was staring us in the face and ignoring the most exciting consequences of the microlensing theory.

John's intervention completely transformed the press release from something that might be of interest to a few specialist journalists and amateur astronomers into a highly newsworthy event, despite the fact that we had comprehensively hedged it with caveats and conditionals. Our press officer decided that it was sufficiently stimulating to be upgraded to a press conference, since it covered a variety of topics of great popular interest: quasars, gravitational lenses, black holes, and the potential to revolutionize our view of the Universe.

On the day of the press conference, there was a good turnout of journalists at the London venue. After I had discussed the main points of the paper, there was an extensive period of questions, many of which went far beyond what I had been talking about. I was very impressed by the astuteness of many of the journalists, who quickly perceived some of the more bizarre consequences of the idea. There was much discussion of the nature of primordial black holes and, strangely, a lot of interest in the implication that we must live in a flat and infinite Universe. After the press conference I retold the story for BBC and Channel 4 News, and agreed to do a piece for *Tomorrow's World*. There were also several radio interviews, and the idea was incorporated in a television documentary on dark matter which was about to be broadcast. All this publicity brought home to me the great popular interest in new and controversial ideas. However, I was to find that this was not an enthusiasm that was encouraged by the scientific establishment.

The television news carried interesting and accurate accounts of the idea and its significance. Newspapers showed a wide range of approaches from the dryly factual to the perceptive, speculative, and sensational. I was nonetheless struck by the essential accuracy of the articles. There are many tales of journalists distorting the statements of people they have interviewed, but I am full of admiration for the intelligent and interesting way in which the story was handled. It is true that some of the headlines were a bit misleading, but I gather that they are added afterwards by a subeditor. My favorite was the *Sun*'s article, enigmatically entitled, "Boffin Predicts End of Universe," but which provided a thought-provoking yet accurate account of the idea.

A number of distinguished astronomers were invited by journalists to comment on my paper. Most quite sensibly adopted a neutral line, acknowledging the importance of the idea if it turned out to be correct, but reserving judgment for the time being. However, two important figures in British astronomy were more outspoken. Professor Michael Rowan-Robinson of Queen Mary College was quoted as saying, "Rubbish,"[8] and Professor Sir Martin Rees of Cambridge said that that he did not believe the idea though he had not read my paper.[9] He also advised at least one journalist to "ignore it."[10] Rees's unfortunate remarks caused merriment among some younger astronomers, but there was a reasonable explanation. I had discussed my idea with Martin Rees several weeks before, and he had drawn my attention to some observations[11] that he considered contradicted my argument. My view was that the observations were not relevant to the situation I was describing, and I had made this point in the paper. Rees knew very well what my line of argument was when he said he did not believe it; the fact that he had not read my paper was irrelevant.

On the whole, I was very pleased with the reception of the paper. I think Stephen Hawking typified the reaction of most cosmologists when he said that if the variability of quasars was caused by small black holes, "this would indeed be a very important discovery." But there were more sinister rumblings emanating from Cambridge that were picked up by *The Sunday Telegraph* and *Scotland on Sunday*. They claimed that some distinguished astronomers felt that an idea which had not been generally accepted by the scientific community should not be publicized. Sir Martin Rees was quoted as saying, "It is very embarrassing if astronomers too often claim to find the secret of the Universe and then it falls flat in a few months. I believe those who made this claim will turn out to be embarrassed."[12] In my opinion, the problem is exactly the opposite. Too often, scientific announcements to the public relate to dull, uncontroversial issues that have finally been tied down, or where the long awaited answer was never in any real doubt. Such stuff is very rarely given much coverage, but when it is, it either perpetuates the myth that science is boring or, because the journalist invariably had to hype it up in order to get editorial acceptance, it ends up being misleading and sometimes even false. Very rarely is the public allowed to glimpse the real debate that underlies every true scientific advance. I think this is because, as we have seen earlier, these conflicts are not normally very clean affairs and certainly do not accord with the image of the scientific process which the establishment likes to foster. Alec Boksenberg, director of the Royal Observatories at the time, expressed a view close to mine when he was quoted as saying about my paper, "Nothing is lost in putting these things out, and I think it is better that the odd person has qualms rather than the research goes unnoticed."[13]

13

A UNIVERSE OF
BLACK HOLES

For doubt can only exist where a question exists, a question only
where an answer exists, and an answer only where something *can
be said*.
    —LUDWIG WITTGENSTEIN, *Tractatus Logico-Philosophicus*

IN THE AFTERMATH of the press coverage there was widespread
interest in my paper from other astronomers, and I was invited to
several conferences to explain my results in more detail. My first
confrontation with the astronomical community was at University
College, London, at a conference hosted by Michael Rowan-
Robinson, who was, as we have seen, a blunt critic of my paper. I had a
chance to chat with him beforehand, and it turned out that although he
was certainly not convinced by my idea, the problem he had with it was
that the black holes would be slow moving, and thus constitute cold
dark matter (CDM). The concept that dark matter is in the form of
CDM had run into some difficulties, and Rowan-Robinson had been at
the forefront of those highlighting objections. If one assumes that the
ripples in the microwave background are associated with cold, slow-
moving particles, and asks how their distribution will change between
then and the present day as they move under the influence of their
mutual gravitational attraction, one finds at first sight that structures
similar to galaxies and galaxy clusters will form. This is in contrast to
the case where dark matter is hot (HDM), in the form of massive
neutrinos, for example, where the result bears little resemblance to the
present-day Universe. Only structures on the very largest scale will

form. The problem that concerned Rowan-Robinson was that if one looks in detail at the CDM picture, one finds worrying departures from what we observe. CDM implies a sequence of events in which small structures form first, and then agglomerate into larger ones. But if one attempts to model the evolution of a CDM Universe, there seems to have been insufficient time to make the largest structures that we see. However, there are a number of ways around this difficulty. For instance, Andy Taylor has produced models consistent with the present state of the Universe by assuming that only 80 percent of dark matter is cold, the remainder being HDM in the form of neutrinos.[1] Primordial black holes would make excellent candidates for the cold component in this "mixed dark matter" model.

I think it is little appreciated that when an astronomer assesses a new result his first thought is, "How does this fit in with my own ideas?" If it supports his work he will almost certainly embrace it and forgive any small imperfections. If it conflicts with his position, he will do his best to find flaws in the argument, and challenge the methodology and any necessary assumptions. Although this flies in the face of the image of objectivity which scientists like to foster, I think we must accept that it is an inevitable corollary of the cultural basis of the scientific process. I do not believe that it is something we should try to hide. It is the stuff of any debate, and we must have confidence that in the long run the better arguments will prevail. However, this attitude of vested interest does become a serious problem in the refereeing process by which papers are assessed for publication. The best referees are capable of putting their own views to one side and judging a paper on its own terms. However, there is hardly any astronomer who cannot tell a tale of inexplicable and irrational opposition to at least one of his papers. I suspect that this is usually no more than the result of a conflict with the referee's own established position.

My talk at University College seemed to go down well. The questions were mainly confined to technical points, but I had no illusions that a more determined attempt to refute my idea would not be mounted. The next conference I attended was at the Rutherford Appleton Laboratory, and was a joint meeting for cosmologists and particle physicists. Sitting in the front row was Martin Rees, whom I had not seen since his dismissive remarks to the press. I again described my results, but with more emphasis on primordial black holes in the hope that the particle physicists would offer some thoughts on their formation. Rees immediately confronted me with a challenge. How was it that the very nearby and luminous quasar 3C 273 could vary like other quasars, when the probability of microlensing at so small a distance was negligible? I was nonplused as I had the firm impression that 3C 273 hardly varied at all.

As soon as I got back to Edinburgh, I checked the records of the variation of 3C 273, which had been monitored for over 20 years. My impression had been correct. Unlike every other quasar of its luminosity, 3C 273 has varied very little, showing only the occasional brief small-amplitude outburst. There was certainly no sign of the long-term variability that I had attributed to microlensing in more distant quasars. This was, in a sense, another blind test of the microlensing idea, and served to increase my confidence that I was on the right track.

Over the next few months I was invited to present my ideas at many institutions, at home and abroad. The enthusiasm for this new approach to the dark matter problem was very encouraging, but I knew there was some way to go before the theory gained general acceptance. There was a lot of very useful feedback and suggestions for new tests from other astronomers, and though many possible objections were put to me, the microlensing picture survived unscathed.

The next major hurdle was to prepare a paper for the 1994 National Astronomy Meeting to be held in Edinburgh. Since the publication of the *Nature* paper, I had been working hard on various new tests of microlensing and wanted to present them in a coherent form. I was also a little concerned that Martin Rees's original objection to the paper had not been fully answered. The observations underlying Rees's point[2] involve changes in brightness of the broad emission lines which, as we have seen, characterize the spectra of quasars. When a light source like a quasar is microlensed, it is only the parts that are sufficiently compact that will be amplified. For a source to be significantly magnified it must be smaller than the Einstein radius of the lens. The emission line region, the clouds of gas where the emission lines in a quasar spectrum originate, is much larger than the Einstein radius of even a stellar-mass object, let alone a Jupiter-mass black hole, and so will not be microlensed to any significant extent. On the other hand, the central part, which includes the accretion disk and is emitting continuum light devoid of lines or other features, is very small and will be microlensed. Thus when a quasar is microlensed, the continuum light will vary behind the constant flux from the emission lines, making them more or less prominent against the background. For massive lenses with large Einstein radii the microlensing magnification can be very large. In this case the light from the emission lines will be completely swamped by the continuum.

The paper to which Rees had referred pointed out that such quasars were hardly ever seen.[3] This puts tight limits on the masses of any large population of microlensing objects. Although this line of argument is quite correct, it no longer applies in the case of Jupiter-mass black holes, where the Einstein radius is not much larger than the diameter of the continuum source associated with a quasar's accretion disk. Extreme magnifications will therefore not occur, and the light curves will have the undulating appearance characteristic of the quasars in field 287. Also, quasars will not be seen where the high magnification of the continuum has washed out the emission lines.

In any event, I was able to take this argument one step further. I tracked down some published observations of the spectra of quasars which had been repeated every few years, making it possible to analyze changes in the emission lines and continuum.[4] These observations clearly showed that for quasars sufficiently distant to be microlensed, long-term variability was associated with changes in the amount of light from the background continuum, while the flux from the emission lines remained constant. Thus, what had started as an objection to the microlensing theory turned out to give it further observational support.

At this time I carried out another important test of microlensing. We have seen that one of the hallmarks of a microlensing event in the halo of our Galaxy is that the light curves are achromatic: they are identical in blue and red passbands. Under most circumstances one would expect a similar situation for quasar microlensing. I had been monitoring the quasars in field 287 in red as well as blue passbands since 1983, and so had an excellent set of observations for this test. There seemed to be no real doubt when I compared the blue and red light curves that in nearly every case they were identical. This was confirmed by a more detailed statistical analysis. Here was another important confirmation of the microlensing theory, as most other mechanisms for quasar variability are not achromatic, and typically involve a larger change of magnitude in the blue than in the red.

The paper that I eventually presented at the Edinburgh meeting consisted of a comprehensive case in support of the microlensing hypothesis. Peter Schneider, who was chairing the session, summarized the current status of the theory. He said that he suspected that many in the audience still found it difficult to believe, but that he had spent six months trying to prove the idea wrong, and failed. It had to be taken seriously.

More recently, at a microlensing conference in Melbourne in 1995, Bill Press provided an overview of the proceedings, emphasizing the significance of dark matter in the form of compact bodies as the basis of future microlensing research. One of the German delegates remarked to me, "Bill has certainly given you a good Press!"

The time had come to write a comprehensive paper[5] describing everything that had been achieved up to that point, and going into the detail that had not been possible in the letter to *Nature,* which was limited to 1500 words. I structured the overall format of the paper to present the microlensing idea as a precisely stated hypothesis, and then subjected it to a series of tests. It would of course have been preferable to have an alternative theory with which to compare it, presumably one accounting for variability on the basis of some process intrinsic to the quasars. No such theory existed in a sufficiently detailed form to make precise predictions, so I found it necessary to speculate as to what such a theory might entail, and then see if this straw man could be knocked down.

The question of the intrinsic variability of quasars is central to the whole issue of whether the microlensing theory is to be believed or not. When the basic details of the accretion disk model for quasars were worked out by Rees and others in the 1970s,[6] the typical timescale of variability which needed to be explained was quite short, perhaps a few months. To get this far was quite an achievement. At that time it was certainly not feasible to attempt any more detailed modeling of light curves, which, in any case, only existed in rudimentary form. Further work on accretion disks over the next 20 years had not really sorted out the problem. There is no easy way to account for intrinsic long-term variability. Typically, in contrast to the observed light curves, accretion disks vary neither achromatically nor symmetrically.

But quasars undoubtedly do vary intrinsically at some level. Seyfert galaxies, which are commonly thought to be a type of low-luminosity quasar, show quite large intrinsic variations. The difficulty in the case of quasars is to sort out what is intrinsic from what appears to be caused by microlensing. We have already seen that the small fluctuations in the quasar 3C 273 must be intrinsic. The observations are still not comprehensive enough to tell us the extent to which they differ from the microlensing variations by, for example, showing a color change. But Seyfert galaxies certainly do show color changes when they vary. This has made me interested in the idea that the amplitude of intrinsic variation might decrease with increasing luminosity. This could be explained if the variability were caused by discrete events not related to the amount of light emitted by the accretion disk; for example an occasional supernova explosion or a star falling into the accretion disk. Such events would be much more obvious against a low-luminosity background, and could thus cause Seyfert galaxies to vary in brightness, whereas they would have little effect on the much more luminous quasars.

One can see the difference between intrinsic variation and the effects of microlensing very clearly in the case of a gravitationally lensed quasar that has been split into two images. We have already met such a system in our discussion of the measurement of the Hubble constant. In the famous quasar pair Q0957+561, one can see many small-scale fluctuations in the light curve of one image which, because of the small difference in the distance traveled by the two light beams, is mirrored about a year later in the light curve of the other.[7] These are clearly intrinsic fluctuations, and have the same overall appearance as those of 3C 273. If one offsets one of the light curves by the time difference, and if all the variations are intrinsic, then one should get an exact match. Thus, if the difference between the two light curves is plotted, it should show no change greater than the measurement error. In practice most of the differences between the small features do indeed disappear, but a large long-term difference between the two light curves becomes apparent. It is generally accepted that this cannot be intrinsic, but is caused by something extrinsic to the quasar that can only be the effect of microlensing.

Rudy Schild, who has spent some 20 years monitoring the quasar pair Q0957+561, maintains that there is a complex pattern of residual variations which must be due to microlensing. In particular, he points to 90-day features that he claims are due to planetary mass bodies.[8] This is exciting stuff, and my own feeling at the moment is that they are very likely to be microlensing events, but the nature of the lenses is still uncertain.

One might think that this provides a conclusive argument in favor of the general microlensing of quasars, but it is not quite as simple as that. Gravitationally lensed quasar images lie on either side of a massive lensing galaxy. One can argue therefore that the light from the quasar has ample opportunity to be microlensed by the stars in the galaxy. This makes the situation untypical of most quasars, and so it cannot be used in a statistical sense to support the microlensing hypothesis for all quasars. However, the observed microlensing effects can only be due to the intervening galaxy if it is composed entirely of small, dark compact bodies. But this would provide spectacular evidence for the nature of dark matter in galaxies.

Either way it would provide confirmation of the idea of dark matter in the form of Jupiter-mass primordial black holes. In my view, the evidence at this stage for the microlensing of all quasars is very persuasive, but perhaps not completely conclusive. At present it is the only viable explanation of the observations. The next step in the project involves several new lines of investigation with the aim of removing the remaining areas of doubt. The question of whether or not quasar variation is caused by microlensing is the crux of the matter. If this could be established, the remaining lines of argument accounting for the nature of dark matter are relatively uncontroversial.

There are diverse new ideas for submitting the microlensing theory to further tests. One approach would involve an extension of the MACHO project[9] to detect Jupiter-mass objects in the halo of our Galaxy. The MACHO project was originally set up to look for brown dwarfs. Objects of this mass, nearly one-tenth that of the Sun, will cause microlensing events of a few weeks' duration, and the program was most sensitive to this regime. Jupiter-mass black holes in our Galaxy would produce events lasting around 2 days. Clearly this is a very difficult time-span, since much of the event will be obscured by daylight. So far, although there are rumors of their detection, no reports of such events have been published. However, the efficiency of detection is estimated to be less than 1 percent, which means that the number of detections would have to be multiplied by over 100 to make up for events that were missed. With such a huge correction factor, I do not think one can read too much into the current lack of results. Another difficulty with this approach is that most of the dark matter in the galactic halo is supposed to lie beyond the Large Magellanic Cloud in an unknown distribution, and so is not in a position to microlens LMC stars. It is therefore difficult to put limits on the amount of matter in the galactic halo in this way. It is also far from clear the extent to which primordial black holes would cluster into the halo of our Galaxy. Recent observations have suggested that it might even be possible that the galactic halo is mostly baryonic, although nonbaryonic material would still need to be present on a larger scale.

A completely different approach to testing the microlensing idea is to improve computer simulations of such events. As we saw in the previous chapter, pioneering work in this field by Peter Schneider and others first suggested the connection between quasar variability and microlensing. However, the early simulations were not designed to model the whole cosmological picture. This, combined with limitations in computer hardware, meant that simplifying assumptions were used to reduce computation time. Edinburgh is the national center for the very fast breed of computers known as parallel processing machines. Unlike most computers, which can only make one calculation at a time, the parallel machines have as many as 256 processors operating simultaneously. There is an overhead in synchronizing all the calculations which, depending on the task the computer has been set, can actually negate any advantage gained from the multiprocessor format. Fortunately, gravitational lensing turns out to be an especially suitable problem for parallel processing, since each processor can be assigned a light ray track as it passes among the lenses. This is known as ray tracing. I am working with Alan Heavens, of the University's astronomy department, and Elspeth Minty, a student from the computer department, to attempt a large-scale and detailed simulation of the effects of microlensing. This project is partly an extension of John Peacock's quick simulations of microlensing in a cosmological context, which, as discussed earlier, he undertook in order to obtain some idea of the lower limit of Ω implied by the fact that every line of sight is being microlensed. We hope to construct close replicas of the observed light curves, and to reproduce any changes of structure with redshift or luminosity. We shall also simulate the effects of changing the mass distribution of the lenses, and their contribution to the critical density.

One of the more straightforward approaches to testing the microlensing idea is to look at quasars along lines of sight where we have reason to believe there is an excess of matter, and hence perhaps an increase in the effects of microlensing on the quasar light curves. There is already some evidence of an excess of quasars behind nearby clusters of galaxies. This is commonly attributed to the macrolensing effect of the mass in the clusters making the quasar images brighter, and thus enabling a larger number of more distant quasars to be seen. It could also be due to microlensing, but work I have done with Liliya Williams of Cambridge suggests that there is no increase in the amplitude of variability for quasars behind clusters. This is not entirely surprising, since the clusters are too near to us for there to be an optimum microlensing effect. Perhaps a better approach would be to look in detail at quasar spectra for the telltale signs of absorption lines, which reveal the presence of galaxies along the line of sight at any distance, and see if those for which there is the most intervening material are the most variable. The most serious problem with this approach is that beyond a certain point it is not clear how adding more lenses will affect quasar light curves. Simulations by Peter Schneider and Joachim Wambsganss suggest a complicated picture in which the depth of lenses is not easily related to the amount of quasar variation.

An ingenious test of the microlensing theory was recently suggested to me by Sjur Refsdal, who pioneered the subject of gravitational lensing when he was a student in the 1960s. His test relies on the well-established fact that the Earth is moving relative to the microwave background at about 400 km/s. The timescale of variation which we see when a quasar is microlensed depends on a combination of the motions of the quasar, the lenses and the Earth relative to the microwave background. A representative value of 600 km/s is usually assumed. Refsdal pointed out that if we look at a quasar sample in the direction of the Earth's motion, the Earth will make no contribution to the combined velocity, whereas if we look in a perpendicular direction the Earth's motion will have the maximum effect. This difference should be measurable by observing quasar samples in different directions and measuring the timescale of variation. There are some serious difficulties to overcome, mainly concerned with ensuring that the two samples are exactly comparable. But, undaunted, Refsdal and I are collaborating to carry out this tricky test.

Most of the evidence bearing on quasar microlensing could be greatly strengthened by a larger sample of quasars monitored over a longer period of time. Since the analysis was done for the *Nature* paper, four more years of data have become available which should help to tighten the measurement of timescale. More important, there is now the possibility that redshifts for all the quasars in the field will soon be obtained. So far, over the last 10 years, over 500 have been measured, leaving some 1200 to be observed. A major new development of the Anglo-Australian Telescope should speedily complete this task. When a redshift is measured with a conventional spectrograph, the image of the quasar is centered on an entrance slit so that redshifts can only be measured one at a time. Over the last 20 years or so, the United Kingdom has pioneered the development of fiber fed spectrographs. In these instruments optical fibers are positioned over the field of view of the telescope at the positions of the objects to be observed. These fibers lead to a spectrograph where they are lined up along the slit, thus allowing many spectra to be observed simultaneously. The technology is highly sophisticated, especially since the whole procedure has been automated. Early versions allowed about 50 spectra to be taken over a field two-thirds of a degree in diameter, but in an ambitious new upgrade the field of view is being increased to 2 degrees, with 400 fibers. Although this development was mainly motivated by the idea of carrying out a huge galaxy survey, it is also ideal for the quasar program.

It has always seemed to me like a missed opportunity that the United Kingdom's astronomy program did not capitalize more effectively on our expertise in wide-field astronomy. In the mid-1980s the possibility of funding for a major new telescope arose. The most obvious choice was to build a conventional 8 meter reflector: an instrument with about twice the light-gathering power of the biggest existing telescopes, but following in the footsteps of major American and European projects. An alternative proposal from Cambridge, championed by Donald Lyndon-Bell, was to build a 6 meter wide-field telescope. This instrument would have a field of view of 5 degrees and at least 1000 fibers. It would be capable of surveying the Universe to unprecedented depths, and would provide a unique and highly efficient approach to some of the most important problems in modern astronomy. It would also have the advantage of capitalizing on an area of expertise in which the United Kingdom led the world. Unfortunately, in my opinion, the decision went in favor of two quarter shares in conventional 8 meter telescopes. I cannot help seeing these monsters as the dinosaurs of twentieth-century astronomy. Although they will undoubtedly produce some interesting and important results, I think the future lies with dedicated, specialized telescopes allied to imaginative projects. There are a number of such facilities already in operation, and in my opinion this is the way to produce the most important and fundamental results about the Universe in a cost effective way.

There are also challenges on the theoretical front. A theory of quasar structure which made testable predictions of the nature of intrinsic variability would be a great step forward. The testing of theories would be made a lot easier if one could find a way of decomposing the light curves into variations that have the properties of microlensing and those which do not. The latter could then be used to constrain theories of intrinsic variation. I have some ideas of how this can be done, but it is not going to be easy. For example, it may be possible to distinguish variations involving color change, and compare their statistical properties with the light curves ascribed to models of intrinsic variation.

Another area that needs further investigation is the formation of Jupiter-mass black holes in the early Universe. The critical question here is the nature of the quark–hadron phase transition that took place after the epoch of inflation. Essentially there are two possibilities, a so called first-order or second-order transition. If the transition is second-order, it would be relatively smooth, and black holes would not be expected to form. In contrast, a first-order transition is accompanied by extensive turbulence and would lead to formation of black holes. The order of the transition depends on the mass of the strange quark, the third lightest of the six quarks. Until recently the order of the transition was very uncertain. Formidable amounts of computer time are required to carry out the calculation, but recent work by a Japanese group[10] now strongly suggests that the mass is in the correct range for a first-order transition, and consequently black hole formation. In this picture David Schramm has argued that Jupiter-mass black holes will form in a catastrophic collapse around top quarks.[11] Most of the matter within each event horizon will collapse into black holes. Thus, these most massive members of the quark family act as seeds for the production of the primordial black holes that permeate the Universe.

I think all this is an interesting demonstration of the way we go about the business of making progress in science. It illustrates the dynamic relationship between theory and evidence, so that it is sometimes almost impossible to tell them apart. My picture of the Universe could not have been conceived had it not been for earlier work by Bill Press, Jim Gunn, Sjur Refsdal, Peter Schneider, and many others. This is quite apart from the fact that its validity depends on the whole theoretical and observational framework of the big-bang model of the Universe, and especially the theory of inflation. The idea only developed with the input of Rachel Webster, Joachim Wambsganss, John Peacock, and Andy Taylor, and its future depends on new investigations by various groups around the world, in some cases undertaken with the specific intention of trying to prove it wrong.

It is too easy, with hindsight, to paint a picture of the birth of a new idea as though it had an inescapable inevitability about it, as though it was just waiting like buried treasure to be discovered, and that the moment of revelation would happen at a particular recognizable instant. Certainly in my case the realization of what was involved dawned very gradually, and was largely due to the insight of others. The relevant data, the observational evidence, had been in my possession for years before Rachel Webster suggested the microlensing interpretation. Arriving at the idea of a Universe of primordial black holes was an uncertain journey which started with an almost casual remark of John Peacock. Now that he has managed to persuade me, it seems obvious and inevitable, but this was certainly not so at the time, and John has yet to be convinced of the validity of the microlensing interpretation of my data.

The task of the scientist is not so much to seek an insight into reality as to try to draw together from the multifarious ideas of his time those which have a bearing on the solution to his problem. His satisfaction lies not in the metaphysical contemplation of truth, but in seeing disparate concepts and observations coming together to provide a coherent picture. The excitement is to see the connection between two apparently unrelated facets of nature. It was in this way that my picture emerged of a Universe composed almost entirely of small black holes. As we look to luminous beacons in its farthest depths, we see them shimmering and flickering as their light bends through a haze of small bodies that constitute the bulk of all matter. The stars and galaxies that fill our view as we survey the depths of the Universe are really just a froth delineating the massive, dark unseen structures beneath.

NOTES

CHAPTER 1   CONSTRUCTING EVEREST

1. Caroline van den Brul, *Perceptions of Science*.

2. Karl Popper, *Conjectures and Refutations* and *Objective Knowledge*.

3. Brian Malpass, *Bluff Your Way in Science*, p. 25.

4. Richard Dawkins, *The Blind Watchmaker*.

5. Steven Weinberg, *Dreams of a Final Theory*, p. 129.

6. Bruno Latour and Steve Woolgar, *Laboratory Life*, p. 237.

7. Weinberg, *Dreams of a Final Theory*, p. 149.

8. The best account of steady-state cosmology is in the final chapters of Fred Hoyle's *Home Is Where the Wind Blows*. The current version of the theory is set out in a paper by Hoyle, Burbidge, and Narlikar, in the June 1993 issue of the *Astrophysical Journal*, entitled "A quasi-steady-state cosmological model with creation of matter."

9. For a comprehensive account of orthodox big-bang cosmology, see Malcolm Longair, "The new astrophysics" in Paul Davies (ed.), *The New Physics*, Ch. 6.

CHAPTER 2   DESCENT IN THE MIST

1. Fred Hoyle, *Home Is Where the Wind Blows*, Ch. 23.

2. Hoyle, *The Intelligent Universe*.

3. Hoyle, *Home Is Where the Wind Blows*, Part 3.

4. For a more detailed description of C (for "creation") fields, see Hoyle, *Home Is Where the Wind Blows*, Ch. 28. See also Hoyle, Burbidge, and Narlikar, "A quasi-steady-state cosmological model with creation of matter."

5. Much of the discussion of Fred Hoyle's views in this chapter is based on Hoyle, *Home Is Where the Wind Blows,* Part 3.

CHAPTER 3 SURVIVAL OF THE WEAKEST

1. Quoted in Timothy Ferris, *Coming of Age in the Milky Way,* p. 177.

2. A. H. Guth, "Inflationary Universe."

3. Fred Hoyle, *Home Is Where the Wind Blows,* Ch. 28.

4. A. A. Penzias and R. W. Wilson, "A measurement of excess antenna temperature."

5. R. H. Dicke et al., "Cosmic black-body radiation."

6. For more information on dissident ideas, see Halton Arp et al. (eds.), *Progress in New Cosmologies,* especially the following articles: Arp, "Fitting theory to observation—from stars to cosmology," p. 1; Eric Lerner, "The case against the big bang," p. 89; John and Thomas Miller, "De Sitter redshift: The old and the new," p. 105; and Victor Clube, "Dark matter, spiral arms and giant comets," p. 187.

7. F. Hoyle and W. A. Fowler, "Nature of strong radio sources."

8. For a personal account of this harrowing experience, see Hoyle, *Home Is Where the Wind Blows,* pp. 409–11.

9. Hoyle, *Home Is Where the Wind Blows,* Ch. 28.

10. John Horgan, "Karl R. Popper, the intellectual warrior," pp. 20–21; Steven Weinberg, *Dreams of a Final Theory,* Ch. 7.

CHAPTER 4 REALITY IN THE DARK

1. According to Microsoft's *Encarta* encyclopedia, positivism is a "system of philosophy based on experience and empirical knowledge of natural phenomena in which metaphysics and theology are regarded as inadequate and imperfect systems of knowledge." The early twentieth-century positivists, known as logical positivists, emphasized the importance of scientific verification. They included the Austrian Ludwig Wittgenstein and the British philosophers Bertrand Russell and G. E. Moore. Wittgenstein's *Tractatus Logico-Philosophicus* decisively influenced the acceptance of the overwhelming importance of empiricism, and the rejection of metaphysical doctrines for their meaninglessness.

2. Steven Weinberg, *Dreams of a Final Theory,* Ch. 7.

3. Richard Dawkins, *The Selfish Gene* and *The Blind Watchmaker.*

4. Fred Hoyle, *The Intelligent Universe.*

5. Hugh Montefiore, *The Probability of God.*

6. Gwynneth Matthews, *Plato's Epistemology.*

7. D. W. Hamlyn, *Aristotle's De Anima.*

8. Steven Weinberg, *Dreams of a Final Theory,* p. 143.

9. Weinberg, *Dreams of a Final Theory,* Ch. 7.

10. Brian Malpass, *Bluff Your Way in Science,* p. 19.

11. Stephen Hawking, *A Brief History of Time,* p. 175. "However, if we do discover a complete theory, it should in time be understandable in broad principle by everyone, not just a few scientists. Then we shall all, philosophers, scientists, and just ordinary people, be able to take part in the discussion of . . . why it is that we and the universe exist. If we find the answer to that, it would be the ultimate triumph of human reason—for then we would know the mind of God." This sentiment is an illustration of Wittgenstein's insight that posing questions about unknowables is meaningless, resulting in just this type of speculative metaphysics.

12. Richard Dawkins, *The Blind Watchmaker,* Ch. 1 and 4.

13. Galileo Galilei, as a Renaissance Aristotelian in defiance of the Platonic rationalism of the academic and theological establishments, argued that we could come to understand reality by observation rather than by intellectual idealism. To the dismay of the Catholic church, he espoused the heliocentric cosmology of Copernicus because it was more consistent with his astronomical observations.

14. Brian Malpass, *Bluff Your Way in Science,* p. 44.

15. Eddington is best known for his recognition and popularization of the implications of Einstein's theory of general relativity, although he failed to explain it. He is also well known for organizing the 1919 eclipse expedition that set out to verify a major prediction of Einstein's theory. A field of stars photographed at night was compared to the same stars photographed during a solar eclipse. Eddington's team claimed that this showed that rays of starlight just grazing the Sun's surface were deflected by the gravitational influence of the Sun's mass by 1.75 seconds of arc. This is exactly what Einstein had predicted. When asked what his response would have been had Eddington's experiment not been consistent with general relativity, the elated Einstein remarked that he would have had to pity God for failing to organize the Universe properly. As it turned out, Eddington's claim was not validated by his data since the margins of error were far too great for any unbiased inferences to be drawn.

16. Karl Popper, *Conjectures and Refutations* and *Objective Knowledge*.

17. John Horgan, "Karl R. Popper, the intellectual warrior."

18. Richard Dawkins, *The Blind Watchmaker*, Ch. 1 and 4.

19. Dawkins, *The Blind Watchmaker*, Ch. 1 and 4.

20. Fred Hoyle, *The Intelligent Universe*.

21. David Hume, *On Human Nature and the Understanding*.

22. Hume, *On Human Nature and the Understanding*, p. 302.

23. Timothy Ferris, *Coming of Age in the Milky Way*, p. 23.

24. Steven Weinberg, *Dreams of a Final Theory*, Ch. 6.

25. Lucien Price, *Dialogues of Alfred North Whitehead*, p. 28.

26. Francis Darwin, *Charles Darwin's Autobiography*, p. 67.

27. Timothy Ferris, *Coming of Age in the Milky Way*, p. 243.

28. Quoted in George Smoot and Keay Davidson, *Wrinkles in Time*, p. 86.

29. Hawking, *A Brief History of Time*, pp. 174–5.

30. In the second century AD, the Egyptian astronomer Claudius Ptolemy extended and refined Aristotle's geocentric model of the Universe. Although Ptolemy removed the aesthetically satisfying symmetry that had commended it to Aristotelians and Platonists alike, it more or less explained the observed planetary motions and was upheld as the greatest guide to the heavens until the Renaissance. However, because of its ungainliness, which disqualified it as a model of Platonic reality, it was regarded as no more than a useful mathematical fiction. Nevertheless, in the spirit of empirical induction, Ptolemy believed that his model reflected what actually existed, and he maintained that if the solution was inelegant, so was the problem. See Ferris: *Coming of Age in the Milky Way*, pp. 28–31.

CHAPTER 5   SCALING THE UNIVERSE

1. This unique glimpse of the distant universe has been made available to the public over the Internet.

2. W. Freedman et al., "Distance to the Virgo Cluster galaxy M100."

3. Brian Malpass, *Bluff Your Way in Science*, p. 59.

CHAPTER 6   THE PRINCIPLE OF MAXIMUM
TRIVIALIZATION

1. Gerard De Vaucouleurs, *The Cosmic Distance Scale*, pp. 11–12.

2. De Vaucouleurs, *The Cosmic Distance Scale*, p. 12.

3. George Smoot and Keay Davidson, *Wrinkles in Time*.

4. P. F. Scott et al., "Measurement of structure in the cosmic background radiation."

5. M. Schmidt, "3C273: A star-like object with large redshift."

6. John and Thomas Millet, "De Sitter redshift: The old and the new," p. 105; and Halton Arp, "Fitting theory to observation—from stars to cosmology," p. 1; in Arp et al. (eds), *Progress in New Cosmologies*.

7. F. Hoyle and W. A. Fowler, "Nature of strong radio sources."

CHAPTER 7   IN THE LAND OF THE BLIND

1. For an account of the Einstein–Bohr argument, see Timothy Ferris, *Coming of Age in the Milky Way*, pp. 286–8.

2. See Chapter 4, note 1.

3. "Schrödinger's cat" is a graphic illustration of the implications of his uncertainty principle which maintains that it is impossible to specify simultaneously the position and momentum of a particle with precision. If a cat is placed in a closed box with two tiny apertures and an electron is fired at the box, either the electron will go through the first opening operating a mechanism to kill the cat or it will go harmlessly into the second opening. But until the box is opened and the contents examined, the cat will be neither dead nor alive but in both states simultaneously. It is only the act of observation that realizes either potential state.

4. Ludwig Wittgenstein, *Tractatus Logico-Philosophicus* and *Philosophical Investigations*.

5. Wittgenstein, *Tractatus Logico-Philosophicus*, line 7.

6. Wittgenstein, *Tractatus Logico-Philosophicus*, line 6.54.

7. Quoted in Abner Shimony: "Conceptual foundations of quantum mechanics," in Paul Davies (ed.), *The New Physics*, p. 394.

8. Shimony, "Conceptual foundations of quantum mechanics," p. 373.

9. Shimony, "Conceptual foundations of quantum mechanics," p. 373.

10. Joseph Ford, "What is chaos that we should be mindful of it?," in Paul Davies (ed.), *The New Physics,* p. 348.

11. Stephen Hawking, "The edge of spacetime," in Paul Davies (ed.), *The New Physics,* Ch. 4.

12. Quoted in Shimony: "Conceptual foundations of quantum mechanics," pp. 392–3.

CHAPTER 8   THE DARK WORLD OF UNCREATED
REALITY

1. For the background to this theory, see Fred Hoyle, *Home Is Where the Wind Blows.*

2. David Layzer, *Cosmogenesis;* and Michael Riordan and David Schramm, *The Shadows of Creation.*

3. Alan Guth and Paul Steinhardt, "The inflationary universe," in Paul Davies (ed.), *The New Physics,* Ch. 3.

4. For a detailed firsthand account of the discovery of ripples in the microwave background, see George Smoot and Keay Davidson, *Wrinkles in Time.*

5. See various articles in Arp et al. (eds), *Progress in New Cosmologies.*

6. On the cover of Smoot and Davidson's book *Wrinkles in Time.*

7. Quoted in the *Sunday Times,* 18 July 1976.

8. Michael Riordan and David Schramm, *The Shadows of Creation,* p. 176.

9. Riordan and Schramm, *The Shadows of Creation,* p. 179.

10. W. H. Press and J. E. Gunn, "Method for detecting a cosmological density of condensed objects."

11. M. Crawford and D. N. Schramm, "Spontaneous generation of density perturbations in the early Universe."

12. Recent theoretical developments have raised severe doubts about the usefulness of axions in solving the dark-matter problem. However, though the necessary experiments are incredibly difficult, there is now some reason to suppose that neutrinos do indeed have a very small mass.

13. Riordan and Schramm, *The Shadows of Creation,* pp. 179–80.

14. For explanations of GUTs, see John Barrow, *Theories of Everything;* Steven Weinberg: *Dreams of a Final Theory;* and David Layzer: *Cosmogenesis.*

15. Hawking, "The edge of spacetime," in Paul Davies (ed.), *The New Physics,* Ch. 4.

CHAPTER 9   DETRITUS

1. Michael Riordan and David Schramm, *The Shadows of Creation,* Ch. 4; Fred Hoyle, *Home Is Where the Wind Blows,* Ch. 20.

2. C. Alcock et al., "Possible gravitational microlensing."

3. Stephen Hawking, *A Brief History of Time,* Ch. 7.

CHAPTER 10   DARK GLASSES

1. W. H. Press and J. E. Gunn, "Method for detecting a cosmological density of condensed objects."

2. M. R. S. Hawkins, "Gravitational microlensing."

CHAPTER 11   COSMIC BEACONS

1. M. R. S. Hawkins, "Three unusual cataclysmic variable stars."

2. Hawkins, "A study of the galactic halo."

3. Hawkins, "Direct evidence for a massive galactic halo."

4. Hawkins, "Variable galactic objects."

5. B. Paczynski, "Gravitational microlensing at large optical depth."

6. C. Alcock et al., "Possible gravitational microlensing."

7. K. C. Sahu, "Stars within the Large Magellanic Cloud."

CHAPTER 12   SEEING THE UNSEEN

1. M. J. Rees, "Black hole models."

2. See e.g. P. Schneider and A. Weiss, "A gravitational lens origin for AGN variability?"

3. M. R. S. Hawkins, "Gravitational microlensing."

4. Hawkins, "Gravitational microlensing."

5. W. H. Press and J. E. Gunn, "Method for detecting a cosmological density of condensed objects."

6. P. Schneider, "Upper bounds on the cosmological density of compact objects."

7. See Riordan and Schramm, *The Shadows of Creation*, pp. 112–14; and Crawford and Schramm, "Spontaneous generation of density perturbations in the early Universe."

8. Quoted in Rob Edwards, "Rocket for star struck astronomers."

9. Quoted in Tim Radford, "Cosmic mystery emerges from black hole."

10. Quoted in Caroline van den Brul, *Perceptions of Science*, p. 13.

11. C. R. Canizares, "Manifestations of a cosmological density of compact objects."

12. Quoted in Rob Edwards, "Rocket for star-struck astronomers."

13. Quoted in Robert Matthews, "Space hype."

CHAPTER 13   A UNIVERSE OF BLACK HOLES

1. A. N. Taylor and M. Rowan-Robinson, "The spectrum of cosmological density fluctuations."

2. C. R. Canizares, "Manifestations of a cosmological density of compact objects."

3. Canizares, "Manifestations of a cosmological density of compact objects."

4. P. M. Gondhalekar, "Ultraviolet spectra."

5. M. R. S. Hawkins, "Dark matter from quasar microlensing."

6. M. J. Rees, "Black hole models."

7. R. E. Schild and R. C. Smith, "Microlensing in the Q0957+561 gravitational mirage."

8. R. E. Schild, "Microlensing variability of the gravitationally lensed quasar Q0957+561 AB."

9. B. Paczynski, "Gravitational microlensing at large optical depth."

10. Y. Iwasaki et al., "Nature of the finite temperature transition in QCD with strange quark."

11. David Schramm, private communication.

GLOSSARY

ABSORPTION LINES   When the light generated by a star passes through surrounding clouds of gas, the atoms or molecules of the gas absorb light of specific colors, producing black bands in the star's spectrum. Each atom or molecule has its own recognizable pattern of absorption lines, and its presence can thus be detected by analyzing the spectrum. See also EMISSION LINES.

ABSTRACT LANGUAGE   A language that refers only to itself and not to the external world. Mathematics is an example of an abstract language.

ABUNDANCE   The relative proportion of each type of atom in the Universe, more correctly called *cosmic abundance*.

ACCRETION DISK   A whirlpool of matter surrounding a black hole or other compact body. Debris in the form of dust and gas falls into the outer parts of the disk, eventually working its way to the center, where it disappears into the compact body. Matter in the accretion disk is accelerated to very high velocities, causing the disk to emit copious amounts of radiation at optical, ultraviolet, and X-ray wavelengths.

ACHROMATIC CORRECTOR PLATE   A large, two-layered glass corrector plate mounted across the aperture of a Schmidt telescope. It corrects for the optical distortions of the telescope across the whole field of view for a wide range of colors.

ADAPTIVE OPTICS   A modern method for making small corrections to the wave-front of the light entering a telescope to compensate for distortions caused by unsteadiness in the Earth's atmosphere. The corrections are normally made by continually distorting a very thin mirror placed in the telescope's light path. Light from a bright star is used to determine what corrections must be made.

AMORPHOUS GALAXY (IRREGULAR GALAXY)   A galaxy with an ill-defined or irregular shape that does not easily fit into galaxy classification schemes.

AMPLITUDE   The range of variation from maximum to minimum, usually measured in magnitudes, in the light output of a variable astronomical object.

ANALYSIS OF LANGUAGE   The modern philosophical pursuit of unraveling the complexity underlying our use of language in order to come to an understanding of the way we really think and our relationship to so-called external reality.

ANDROMEDA GALAXY   The nearest large spiral galaxy, and the largest member of the Local Group of galaxies.

ANTHROPIC PRINCIPLE   The idea that the Universe is the way it is because if it were any different we would not be here to observe it. In other words, the Universe evolved to suit us.

APPARENT MAGNITUDE   See MAGNITUDE.

ARGUMENT FROM DESIGN   An argument for the existence of God based on the premise that anything that exhibits complexity, order, and design must be the result of purposeful creation, and could not have come into being simply of its own volition. This argument usually takes the form of a catalogue of nature's marvels, accompanied by expressions of personal incredulity at the thought that such intricacy and coherence could have been the result of mere chance. Statistical arguments are frequently employed to lend spurious weight to such incredulity.

ARGUMENT FROM PERSONAL INCREDULITY   Like the argument from design, this is not really an argument but simply a refusal to consider any view that conflicts with an entrenched position or personal prejudice. As Richard Dawkins points out, it is fairly easy to identify because it is usually accompanied by phrases such as "I cannot believe that . . . ," "it is unlikely that . . ." and so on. Another characteristic of an argument from personal incredulity is a heavy reliance on authority, tradition, and precedent.

ARROW OF TIME   In the context of physical laws at the microscopic level being symmetrical with regard to time, the concept of the arrow of time is an attempt to understand why time is clearly asymmetrical and unidirectional—in other words, why it only runs forward.

ARTIFICIAL LANGUAGE   An abstract or pseudolanguage such as a computer language, mathematics, formal logic, or the set of chess moves.

ASTROPHYSICS   The application of the laws of physics, chemistry, and the other physical sciences to astronomy.

ATOMIC MATTER   See BARYONIC MATTER.

AUTOCORRELATION FUNCTION   A mathematical description of the extent to which various timescales dominate the variation of a quantity, such as light.

AXION   A type of heavy, slow-moving subatomic particle postulated in some modern theories of particle physics. Although they would be virtually impossible to detect, axions have until recently been considered as plausible candidates for dark matter.

BARYON   A nuclear particle, such as a proton or neutron, which is built from three quarks.

BARYONIC MATTER   Matter composed of baryons; atomic matter.

BIG-BANG THEORY   The theory that the Universe began about 10 to 20 billion years ago in a state of enormous temperature and density, and has since expanded and cooled to its present state.

BILLION   A thousand million.

BLACK-BODY CURVE   An asymmetrical curve showing the relation between the power of the radiation emitted by a black body across a spread of wavelengths. A black body is a hypothetical object, though the radiation from many astronomical objects, including stars, resembles a black-body curve. A black body absorbs all radiation incident upon it, which it then re-emits in a characteristic distribution that depends solely upon its temperature. The distribution will peak at a specific wavelength (color), which becomes redder as the body gets cooler. The cosmic background radiation, which is very cold at about 2.7 degrees above absolute zero, shows a perfect black-body curve with a peak in the microwave region, beyond the infrared.

BLACK HOLE   A region where matter is so condensed, and hence gravity so strong, that nothing—not even light—can escape. The boundary of this region is known as an EVENT HORIZON. Black holes can form as the result of the catastrophic collapse of matter. This matter can be in the form of a massive star. On a much larger scale, where conditions allow, thousands or millions of stars can coalesce to form a black hole. Two situations where black holes are expected to form are when a star more than 4 times the mass of the Sun reaches the end of its life and is unable to withstand its own intense gravitational field, and in the center of large galaxies, where up to 100 million stars may accrete into a massive black hole. Another circumstance when black holes could form is in the first microseconds of the big bang, when they could be much smaller than the limit of four solar masses. These are known as PRIMORDIAL BLACK HOLES.

BROWN DWARF   A very low-mass star, less than 8 percent of the mass of the Sun. This mass is insufficient to trigger the thermonuclear reactions that power true stars and prevent them from collapsing. All the energy of brown dwarfs is derived from the initial energy of their collapse. They are very cool, and hence strong infrared emitters. It is believed that they could make an important contribution to the dark matter within galaxies.

195

CATACLYSMIC VARIABLE   In the later stages of a star's life its structure can become unstable in a number of ways, some of which lead to periodic explosions or outbursts. Stars in this state are known as cataclysmic variables. The most extreme examples of this type of behavior are SUPERNOVA explosions, in which an entire star is disrupted.

CAUSALITY   The basic assumption of classical physics, and indeed of commonsense thinking: that objects and events are interconnected and that every new situation must have been caused by a previous state so nothing can come into being from nothing.

CEPHEID VARIABLE   During the later period of a star's life, instabilities form in its outer atmosphere, causing it to pulsate regularly and thus change in brightness in a characteristic way. There are a number of classes of such stars, the most luminous of which are known as Cepheid variables. They have the useful property that their period of variation, which is easily determined by observation, can be related to their luminosity. They can thus be used as "standard candles" for the purpose of measuring distances.

C-FIELD   A field postulated by Fred Hoyle in his revised steady-state theory to govern the continuous creation of matter. The C-field can be incorporated in the equations of general relativity, and as the Universe expands it allows the density to be maintained by the creation of new matter.

CHAOS   Unpredictable and apparently random behavior in a system or structure. The study of chaos has become a major part of modern mathematics, seemingly revealing very simple fundamental laws.

CHARGE COUPLED DEVICE (CCD)   A rectangular array of microscopic detectors that can be used in place of photographic film to record an image. Typical modern CCDs consist of 1 million detectors arranged in 1000 by 1000 array of pixels (picture elements). Their advantages over film and other, earlier detectors are their very high sensitivity (nearly 100 percent of photons can be detected) and their simple linear response to the light incident upon them. They cannot yet compete with photography for survey work because they can cover only a very small area of sky.

CLASSICAL PHYSICS   Physics that does not incorporate or take into account quantum mechanical effects. Typically, it still provides an accurate description of nature on all but the smallest scales.

CLOSED   Describing a system that is self-contained, completely detached from any external environment, so that no interchange of energy or mass or any communication with that environment can take place.

CLOSED UNIVERSE   A universe of finite extent. Such a universe will in principle have a measurable volume, but will not necessarily be bounded. A two-dimensional example is the surface of a sphere, which has a measurable area but no edge.

COLD DARK MATTER (CDM)   It is widely believed that around 90 percent of the Universe is in a dark and so far undetected form. Computer simulations of the evolution of the Universe since DECOUPLING suggest that most of this matter must consist of slow-moving (i.e., cold) material. A Universe structured from such material is known as a cold, dark matter Universe.

COMMON LANGUAGE   The class of all natural languages such as English, Sanskrit, or Swahili.

COMPACT BODY (CONDENSED BODY)   Material in the form of a discrete entity such as an asteroid, a planet, or a star, as opposed to free-moving elementary particles, atoms or molecules.

COMPANION STAR   The less massive component of a binary star system (two stars revolving about their common center of mass).

CONTINGENCY   Whatever happens to be the case, as opposed to what has to be the case, or NECESSITY.

COSMIC BACKGROUND RADIATION   Microwave radiation that permeates the Universe. This radiation is highly isotropic (the same in all directions) and has a wavelength distribution very close to that of a black body at a temperature of 2.7 degrees above absolute zero. It is widely believed to have originated when the Universe became transparent about 100,000 years after its creation. It has been cooling ever since, and is thus seen as the afterglow of the fireball of the big bang.

COSMIC (OR COSMOLOGICAL) PARAMETERS   General relativity provides a framework for describing an enormous set of possible universes. Any particular universe is characterized by a number of general properties such as isotropy and homogeneity, together with parameters describing its size, mass density, and so on. These are known as cosmic (or cosmological) parameters.

COSMIC STRINGS   Extremely massive threadlike objects, whipping through the Universe, predicted by some GRAND UNIFIED THEORIES. They are thought of as cracks in space-time, or as the afterbirth of PHASE TRANSITIONS when the early Universe condensed from one state into another.

COSMOLOGICAL CONSTANT   A cosmic parameter which, with gravity, governs the rate of expansion or contraction of the Universe. It was first introduced by Einstein to balance gravity exactly and produce a static Universe. More recently, it has been employed to allow more freedom in reconciling other cosmic parameters with observation.

COSMOLOGICAL PRINCIPLE   The basic assumption that the Earth is not located in any special position in the Universe, and that the large-scale features of the Universe would therefore look much the same to any other observer as they do to us. This idea has been extended by proponents of the steady-state theory to the *perfect cosmological principle:* that our position in time as well as in space is not special in any way, and that the Universe therefore looks the same in the past as it does now and will do in the future.

COSMOS   An automated plate-measuring machine located at Edinburgh which rapidly scans photographic plates, detecting stars and galaxies and recording a number of parameters describing them. It can also provide a two-dimensional scan that can be reconstituted by computer to provide a digital picture. The output can be used for large-scale surveys and statistical studies of a wide variety of astronomical phenomena.

CREATIONISM   The belief that the Universe was created by a divinity, and that life did not arise through Darwinian evolution but came into being all at once, more or less just as it is now.

CRITICAL DENSITY   In the big-bang picture of the Universe there are two basic possibilities. If the mass density is above a certain critical value, the Universe will be finite in extent, and after an initial phase of expansion it will succumb to the force of gravity and collapse back to a singularity. If the density is less than the critical value, the Universe will be infinite, and gravity will be insufficient to prevent it from expanding for ever. The theory of INFLATION and other considerations suggest that the actual density of the Universe is in fact very close to the critical density, which would mean that the Universe will expand indefinitely, but ever more slowly. The cosmological parameter describing the density of the universe is Ω (OMEGA). For an empty universe $\Omega = 0$, for the critical density $\Omega = 1$, for an open universe Ω is less than one, and for a closed universe Ω is greater than 1.

DARK MATTER   Observations over the last 50 years or so strongly suggest that we have yet to see most of the matter in the Universe. This material does not appear to emit light or any other form of radiation, and has come to be known as dark matter or *missing mass.*

DARWINISM   The theory that species arise through the *natural selection* of random mutations that better fit the changing conditions of their general environment.

DECELERATION PARAMETER   The density of the Universe determines the rate at which it expands. The deceleration resulting from the pull of gravity is measured in terms of the deceleration parameter, q_0. In a big-bang Universe there is a simple relation between q_0 and the density parameter Ω, such that $2q_0 = \Omega$.

DECOUPLING   In the early Universe there was a constant interchange between photons and particles of baryonic matter. This made the Universe opaque as photons could only travel a short distance before interacting with such matter. Equally, baryonic matter was prevented from condensing by the continual interference from the sea of photons. After about 100,000 years the Universe cooled to a point where such interchanges no longer took place, and photons became free to travel indefinitely long distances. This is known as decoupling. Thus the Universe became transparent, and baryonic matter was free to start participating in the formation of the structures that we see today.

DEGENERATE STATE   A state in which the atoms of a gravitationally collapsed star or other body are crushed so that the electrons and nuclei become tightly packed together in a lattice. Matter in this form is extremely dense.

DETERMINISM   The philosophy that all events are completely determined by prior events, by natural laws, or the will of a divinity.

DISCREPANT OBSERVATION   An observation that appears to conflict with widely believed theoretical concepts.

DISCREPANT REDSHIFT   A redshift that cannot be attributed to the expansion of the Universe. A discrepant redshift may be suspected when two objects with widely different redshifts lie close together on the sky, and statistical arguments, or the observation of filaments of material apparently connecting them, suggest that they are in real physical proximity. The reality of all the discrepant redshifts so far reported is strongly contested by most astronomers.

DOPPLER SHIFT   A change in the apparent wavelength of any form of radiation or sound emitted by a moving body. In an expanding Universe, this has the effect of increasing the wavelength of light from the more distant and therefore more rapidly receding galaxies, thus shifting their colors toward the red. This manifestation of the Doppler effect is known as REDSHIFT.

DOUBLE QUASAR   Among the many thousands of quasars so far discovered, a small number appear as pairs of almost identical images separated by a few seconds of arc. Such close pairings cannot easily be accounted for by chance. One explanation is that the quasar image has been split in two by the gravitational lensing effect of a massive galaxy lying on our line of sight to the quasar. Such systems can tell us much about the lensing galaxy and about the nature of apparent differences between the two quasar images.

DWARF GALAXY   A galaxy at the lower end of the size range, containing between a few hundred thousand and a few million stars. Dwarf galaxies are by far the most common type, although their small size makes them difficult to see at great distances. The Local Group of galaxies, to which our own Galaxy belongs, contains over 30 members, many of them dwarf galaxies.

The Magellanic Clouds (see LARGE MAGELLANIC CLOUD) are the most prominent local dwarf galaxies.

ECLIPSE   In a binary star system viewed edge-on, each star will periodically totally or partially obscure the other as it crosses the line of sight in the course of its orbit around the other. These events are known as eclipses. A similar effect occurs when the Moon crosses the line of sight from the Earth to the Sun, causing a solar eclipse.

EINSTEIN CROSS   A dramatic example of gravitational lensing, where a distant quasar is split into four separate images by the presence of a massive galaxy along the line of sight.

EINSTEIN RADIUS   The distance out to which a compact body will have a significant gravitational lensing effect on the light from a more distant object.

ELECTROMAGNETIC RADIATION   Radiation that travels at the speed of light and can pass through a vacuum. It can be treated as either waves of various wavelengths, propagated by rapidly alternating electric and magnetic fields, or a stream of particles or "photons" of different energies. The range of electromagnetic radiation across all wavelengths is called the electromagnetic spectrum. Light is the best-known form of electromagnetic radiation. At shorter wavelengths (higher photon energies) than light are ultraviolet radiation, X-rays, and gamma rays; at longer wavelengths (lower energies) are infrared radiation, microwaves, and radio waves.

ELEMENTARY PARTICLES   Indivisible particles such as electrons, neutrinos, and the most fundamental building blocks of atomic matter, the quarks.

ELLIPTICAL GALAXY   One of the main categories of galaxy types. In telescopes they appear smooth and featureless, with elliptical outlines. They contain mostly old stars and are largely free of gas. They tend to be found near the center of large galaxy clusters.

EMISSION LINES   The light emitted by atoms in the hot gas of stars or galaxies is at certain wavelengths, and shows up as bright, narrow lines in the star's spectrum. Each element has a characteristic pattern of lines representing different colors, and by analyzing a body's spectrum one can identify particular elements as being present in the body. See also ABSORPTION LINES.

EMPIRICISM   The doctrine which affirms that all knowledge is based on experience. It denies the possibility of spontaneous ideas or a priori thought, and is in direct opposition to RATIONALISM.

ENTROPY   A property of a physical system which can be thought of as the degree of disorder in the system. The *second law of thermodynamics* states that in a closed system entropy can only increase.

EPISTEMOLOGY   Theories of knowledge. See EMPIRICISM, RATIONALISM, REALISM, IDEALISM.

EUCLIDEAN GEOMETRY   The most familiar form of geometry, which describes figures drawn on a flat, infinite plane. It may be contrasted with the more complicated geometries describing figures drawn on curved surfaces.

EVENT HORIZON   The distance that light has traveled since the creation of the Universe. It is thus the farthest distance to which it is possible for us to see, and also the boundary of the region within which we can have any causal contact with any other object. Another way of regarding the event horizon is as an insuperable space-time barrier. A black hole is bounded by its own event horizon, beyond which no light or any other form of information can intrude into the observable Universe. Also, two regions of the Universe separated from each other by a distance which, when measured in light years, is greater than the number of years for which the Universe has been in existence, can be said to be separated by an event horizon. Since two such regions occupy different *horizon volumes,* they are not causally connected.

EXPANSION OF THE UNIVERSE   The phenomenon in which galaxies are in general receding from one another. The velocity of recession increases in proportion to distance.

FALSIFIABILITY   The property of a theory that makes predictions which are sufficiently precise for them to be shown to be inconsistent with observations. In other words, such a theory is capable of being proved wrong by experiment. Under the influence of Karl Popper, this is generally considered the minimum requirement for an acceptable scientific theory.

FALSIFICATION   The process of attempting to find evidence that the predictions of a theory are not fulfilled, thus demonstrating that the theory is wrong.

FIELD   A physical quantity, such as a magnetic field or gravitational field, which varies from point to point in space as opposed to a particle, which exists at only one point at a time.

FLAT UNIVERSE   A special case in the family of universes with curved space-time. In a flat universe, the geometry is EUCLIDEAN GEOMETRY and the value of the density parameter Ω is unity, the CRITICAL DENSITY.

FLATNESS PROBLEM   A universe created with an approximately flat geometry will tend as it evolves to rapidly diverge to an open or closed state. It is only if the universe starts almost unimaginably close to perfect flatness that it can remain so over time. After 10–15 billion years, we find ourselves today in a Universe that is still very close to being flat. The riddle is why the Universe was born so close to perfect flatness.

GALACTIC BULGE   The swelling of stars at the center of a spiral galaxy, also known as the *galactic nucleus*.

GALACTIC DISK   The thin disk of young stars and gas surrounding the central bulge of a spiral galaxy, consisting of a number of distinct spiral arms.

GALACTIC HALO   A diffuse, more or less spherical cloud of old stars surrounding the disk and central bulge of a spiral galaxy. The halo is the oldest part of a galaxy, and galactic halos are believed to contain a substantial amount of dark matter which has yet to be identified.

GALAXY   A vast structure of stars and gas containing between about 10^6 and 10^{12} times the mass of the Sun. There are three major types: spiral, elliptical, and amorphous (or irregular).

GALAXY   The galaxy in which the Sun is located, given a capital G to distinguish it from others; sometimes called the Milky Way Galaxy.

GAMMA RAYS   Electromagnetic waves of very short wavelength, produced in radioactive decay or by collisions between elementary particles.

GENERAL RELATIVITY   See RELATIVITY.

GIANT MOLECULAR CLOUDS   The disk of a spiral galaxy contains vast clouds of gas, much of it cold and in molecular form. It is from these giant molecular clouds that stars are formed.

GLOBULAR CLUSTER   A spherical cluster of up to 10 million stars orbiting in the halo of a galaxy. The stars in globular clusters are among the oldest in a galaxy, and their ages can be used to determine the age of the galaxy itself.

GRAND UNIFIED THEORY (GUT)   A mathematical scheme in which fundamental forces of nature are unified in a consistent description. The strongest of these forces is the strong nuclear force, which binds together quarks to form protons and neutrons, and holds the protons and neutrons together in the nuclei of atoms. The second strongest is the electromagnetic force, which operates between particles with electric charge and so holds atoms together, and the third strongest is the weak nuclear force, which is involved in radioactivity. These last two have been unified in the so-called electroweak theory. A GUT attempts to unify the electroweak and strong nuclear forces. The gravitational force, the fourth fundamental force, is by far the weakest and has not yet been unified with the others.

GRAVITATIONAL FIELD   Massive objects attract each other with a gravitational force. The size of this force at any point in space is described by the gravitational field.

GRAVITATIONAL FORCE   Fundamental force of nature, generated by all particles that possess mass. The strength of the gravitational force is proportional to the mass of the object and decreases with the square of the distance from the object.

GRAVITATIONAL LENSING   The gravitational fields of massive bodies can bend light. When such a body lies along our line of sight to a distant light source, it can act as a lens, magnifying and distorting the image of the distant source.

HADRON   Any particle made up of quarks or antiquarks, such as protons, neutrons, antiprotons, antineutrons, and mesons. Electrons, muons, and neutrinos are examples of particles that are not hadrons.

HAWKING RADIATION   The radiation produced at the surface of a black hole as a result of quantum mechanical effects. By emitting Hawking radiation, sufficiently tiny black holes can eventually evaporate completely.

HELIOCENTRIC UNIVERSE   The Universe with the Sun at the center, in contrast to the Earth-centered *geocentric Universe,* which was the standard model until after Copernicus.

HELIUM   A light element whose nucleus consists of two protons and two neutrons. It was synthesized during the era of nucleosynthesis, and is also created in stars when four hydrogen atoms fuse into a helium nucleus and release energy. This reaction is a principal source of the energy that powers stars.

HIDDEN VARIABLES THEORY   A theory which denies that the quantum state of a physical system is a complete specification of the system.

HOMOGENEITY   Uniformity in all directions.

HORIZON PROBLEM   The finite speed of light means that two separate points in the Universe will be able to "see" each other only after there has been sufficient time for light to travel between them. Thus to start with every point in the Universe will evolve independently of all other points. The horizon problem is the mystery of how separate parts of the early Universe, which must have been causally disconnected because there could have been no communication between them, could nevertheless have almost identical properties. See also EVENT HORIZON.

HORIZON VOLUME   See EVENT HORIZON.

HOT DARK MATTER (HDM)   It is widely believed that some 90 percent of the mass of the Universe has yet to be identified. This dark matter may be cold, slow-moving particles (see COLD DARK MATTER) or hot particles moving at high velocities, such as neutrinos. Computer simulations of the evolution of the Universe suggest that HDM is not a major component of the dark matter.

HUBBLE CONSTANT (H_0)   Distant galaxies are receding from us (and, by inference from one another) with a velocity v proportional to their distance *D*. The Hubble constant relates these two quantities thus: $v = H_0 D$.. This relationship is called *Hubble's law*.

HYADES   The nearest cluster of stars to the Sun.

HYDROGEN   The lightest element, consisting of one proton orbited by one electron. It is the most abundant element in the Universe, and fuels the stars.

IDEALISM   The doctrine, originating with Plato, which regards "universals" such as time or redness as self-subsisting metaphysical entities. In the idealist worldview, the material objects we experience are no more than vague and unreal shadows of the ultimate reality of the realm of ideals. Like bad vacation snapshots, such material objects are no more than poor representations of the real thing. See also REALISM, RATIONALISM.

INDETERMINISM   Since the position and trajectory of a particle cannot both be known with precision (see UNCERTAINTY PRINCIPLE), this led to the precept of quantum physics that the information we obtain about natural events is conditioned by the nature of the questions we choose to ask.

INDUCTION   Natural or commonsense thinking, in which the conclusion, though suggested by the premises and consistent with them, does not necessarily follow from them. This is the basis of empirical systems of knowledge. So, for instance (leaving aside subsequent justifications), the Ancient Egyptian belief that the star Sirius causes the Nile to flood is as valid an empirical idea as our belief that the Moon and Sun cause tides. Neither belief is the result of deduction or formal logic, but of induction.

INFLATION   The epoch of extraordinarily rapid expansion of the Universe within the first 10^{-30} second after the big bang, when it increased its diameter by a factor of at least 10^{25}.

INFRARED   Electromagnetic radiation with a wavelength between about 0.001 and 0.3 millimeters.

INITIAL CONDITIONS   The initial state of a system before it evolves with time.

INTRINSIC LUMINOSITY   See LUMINOSITY.

ISOTOPE   One of a number of atomic varieties of a particular chemical element, differing in the number of neutrons present in the nuclei of the element's atoms. All isotopes of a particular element have the same number of protons in their atomic nuclei, and display the same chemical properties.

ISOTROPY   The quality of being the same in all directions.

JUPITER   The largest of the planets in our Solar System, with a mass around one-thousandth that of the Sun and about 1000 times that of the Earth.

KINEMATICS   The study of the motions of bodies such as stars in a galaxy, or of galaxies within galaxy clusters.

LANGUAGE, THEORIES OF   See LINGUISTICS.

LARGE MAGELLANIC CLOUD (LMC)   The largest of the dwarf galaxies in the Local Group, the galaxy cluster to which our Galaxy belongs. It has a primitive spiral structure and contains about 10 million stars.

LARGE-SCALE STRUCTURE   The structure of the Universe on the largest scales we can observe, typically larger than the scale of galaxy clusters.

LIGHT   Electromagnetic radiation with wavelengths of, or close to, those detectable by the eye.

LIGHT CURVE   A plot of the change of brightness with time for a variable star or other astronomical body.

LIGHT ELEMENT   An element, such as helium or lithium, with a small number of protons and neutrons in its nucleus.

LIGHT-YEAR   The distance traveled by light in one year.

LINGUISTICS   A twentieth-century philosophical movement including LOGICAL POSITIVISM. It holds that the proper activity of philosophy is to clarify concepts by analyzing language as it is ordinarily used, thus resolving philosophical problems that are, arguably, the result of linguistic confusion.

LITHIUM   One of the light elements created during the era of nucleosynthesis.

LOGICAL POSITIVISM   Strongly influenced by Wittgenstein's *Tractatus Logico-Philosophicus,* logical positivists maintain that the only factually meaningful propositions are those of science and other empirically derived methodologies which rely on ordinary causal or inductive reasoning. Metaphysical, theological, and ethical sentences are considered to be factually meaningless, as are purely deductive propositions.

LOW-MASS STAR   Conventionally, a star whose mass is barely sufficient to support hydrogen burning, as opposed to a brown dwarf, which is not massive enough to burn hydrogen at all. The dividing line between the two categories is about 8 percent of the Sun's mass.

LUMINOSITY   The intrinsic amount of light emitted by a luminous body such as a star or galaxy. This is in contrast to the apparent brightness, which is the amount of light detected by the observer.

LUMINOSITY FUNCTIONS   Relations describing the relative numbers of stars in successive intervals of luminosity in a given sample of space.

MACHO   See MASSIVE ASTROPHYSICAL COMPACT HALO OBJECT.

MACROLENSING   Gravitational lensing on the large scale, where no changes are apparent over very long periods of time. Typical examples are the multiple imaging of quasars by galaxies, and the spinning of the images of distant galaxies into arcs by massive galaxy clusters.

MAGNETIC MONOPOLE   A hypothetical, extremely massive particle. Unlike all known magnets, which are bipolar, a magnetic monopole carries an isolated north or south magnetic pole.

MAGNITUDE   A measure of brightness for astronomical objects. It is a logarithmic scale such that five magnitudes corresponds to a change in brightness by a factor of 100. The larger the magnitude, the fainter the object. *Apparent magnitude* is a measure of the light detected from an astronomical body. It may be defined formally as -2.5 times the logarithm of the light flux plus a constant. The *absolute magnitude* of a body is a measure of its intrinsic brightness, and is defined as the apparent magnitude it would have if it were located a distance of 10 parsecs from the Earth.

MALMQUIST BIAS   The selection effect by which one is inclined to infer that brighter objects at greater distances are more numerous than they really are. This is because they are more visible than fainter objects.

MANY-WORLDS COSMOLOGY (MULTIVERSE THEORY, PARALLEL UNIVERSES)   "Theories of everything" (TOEs) purport to unify gravity with the other three fundamental forces to describe the Universe as it was in the first 10^{-43} second of its existence. In all TOEs, parallel universes can arise. These "shadow universes" occupy the same physical space as our Universe but, with the exception of gravity, they have completely different forces and particles. An important consequence of this idea is that our Universe would feel the gravitational forces of these shadow universes and would therefore be causally connected to them. So, for instance, the problem of the missing mass could be explained as the gravitational influence of the total mass of all these other universes. A major difficulty with this cosmology is that there is no limit to the number of possible parallel universes. There could in principle be infinitely many, resulting in an infinitely strong gravitational force, which in effect would make our Universe and all the other parallel universes infinitely dense.

MASS FUNCTIONS   Relations describing the relative numbers of stars in successive intervals of mass in a given sample of space. Typically, the smaller the mass, the greater the number of stars.

MASSIVE ASTROPHYSICAL COMPACT HALO OBJECT (MACHO)   The halo of our Galaxy contains a large proportion of matter in a form not yet identified. One possibility is that the material is in the form of compact bodies. These have come to be known as massive compact halo objects.

MEGAPARSEC   A unit for large cosmological distances, a megaparsec (1 million parsecs) is about 3 million light-years, and is roughly the distance to the nearest large spiral galaxy, M31 in Andromeda.

METALANGUAGE, METALINGUISTICS   Words like 'noun' and 'preposition' belong to a metalanguage—a language about language. Logical positivists argue that artificial languages like mathematics are also metalanguages, or 'pseudolanguages,' as are any deductive or rationalist systems, including philosophy itself. Metalanguages are essentially meaningless because they do not deal with matters of fact and existence; only natural languages, which are founded on inductive reasoning, can do this.

METALLICITY   The composition of stars may be broadly divided into three components: hydrogen, helium, and all other elements put together. In astrophysics, all these other elements are referred to as *metals*. The ratio of metals to hydrogen and helium in a star is known as its metallicity, and is generally stated in terms of the metallicity of the Sun. Old stars have low metallicities, as they were formed before the interstellar medium was enriched by new metals fabricated in supernovae explosions.

METAPHYSICS   In traditional philosophy, metaphysics applied to all reality; it was concerned with the nature of ultimate reality, and so included physics. However, the logical positivists disputed the value of metaphysical thinking, claiming that it had no factual meaning, and under their influence the term has entered popular usage as referring to matters transcending material reality. This is the sense in which it is used in this book—standing in opposition to physics.

MICROLENS   A body that gravitationally microlenses a light source.

MICROLENSING   A small body crossing our line of sight to a distant light source causes a change in brightness due to gravitational lensing. For the passage of a single body there is a symmetrical brightening and fading. If a number of bodies simultaneously cross the line of sight, a complicated pattern of fluctuations will be seen.

MICROLENSING HYPOTHESIS   The hypothesis that all quasars are being microlensed.

MICROWAVE BACKGROUND RADIATION   See COSMIC BACKGROUND RADIATION.

MICROWAVES   Electromagnetic waves with a wavelength of around 1 millimeter.

MILKY WAY   See GALAXY.

MISSING MASS   See DARK MATTER.

MONOPOLE   See MAGNETIC MONOPOLE.

MULTIVERSE   See MANY WORLDS COSMOLOGY.

MYSTIFICATION   Making matters so opaque and confusing that the un-initiated are left with the impression that it is too profound and elevated to comprehend.

NATURAL LANGUAGE   See COMMON LANGUAGE; compare ARTIFICIAL LANGUAGE.

NATURAL LAW   An idea that has such universal applicability that any violation of it is deemed to be almost impossible. Oddly, most of Newton's ideas are called laws, whereas those of many other scientists, including Einstein, are still referred to merely as theories. Compare CONTINGENCY.

NATURAL SELECTION   See DARWINISM.

NEBULA   General term for various objects appearing through the telescope as diffuse patches of light. When first seen, their nature was unclear. We now know that some of them are other galaxies, while others are star clusters and gas clouds within our own Galaxy.

NECESSITY   The opposite of CONTINGENCY; the way things have to be as opposed to the way they just happen to be. Logical positivists maintain that there is no such thing as natural necessity: everything is contingent, even so-called natural laws. The only necessity is the logical necessity expressed in the pseudopropositions of METALANGUAGES, which are tautological and consequently meaningless.

NEUTRINO   An electrically neutral, extremely light, and possibly massless elementary particle. Neutrinos interact only very rarely with other forms of atomic matter.

NEUTRON   An uncharged elementary particle, similar in mass to the proton. Protons and neutrons are present in the nuclei of atoms in roughly equal numbers.

NEWTONIAN PHYSICS   Physics based on Newton's laws of motion and gravitation, which govern all local phenomena.

NUCLEAR FUSION   The collision and coalescence of two atomic nuclei to form a single heavier nucleus.

NUCLEOSYNTHESIS   Nuclear processes by which chemical elements are synthesized, principally in the central regions of stars where temperatures are sufficiently high. Primordial nucleosynthesis, in which the lightest elements were synthesized, took place in the hot early phases of the big bang, the so-called nucleosynthesis era. Explosive nucleosynthesis, in which heavier elements are formed, takes place during supernova outbursts.

NUCLEUS   The core of an atom, consisting of protons and neutrons held together by the strong nuclear force.

OBSERVATIONAL COSMOLOGY   The use of observation to determine cosmic parameters, with the aim of distinguishing between rival cosmologies.

OCCAM'S RAZOR   The principle that, when one is faced with a choice between two rival theories or explanations, the simplest should always be preferred. It is sometimes called the *principle of parsimony*.

OMEGA (Ω)   The smoothed out density of the Universe may be expressed as a fraction of the critical density. This fraction is known as Ω, the cosmological density parameter. In a critical or flat Universe, $\Omega = 1$. See also CRITICAL DENSITY.

ONTOLOGY   Ideas of reality, of what there is and the way things are in the objective world.

OPEN   Describing any system that communicates with an external environment, for example by the exchange of energy or matter.

OPEN UNIVERSE   A universe that is unbounded and infinite in extent; where the density parameter Ω is less than unity and the geometry is such that the sum of the angles of a triangle would be less than 180°.

PARADOX   A self-contradictory proposition. Exposing paradoxes in belief systems is one of the most useful devices for pointing out flaws in the data or reasoning that led to such systems.

PARALLAX   A measure of the distance of a star from the Sun. It actually refers to the displacement, relative to the background of more distant stars as viewed from the Earth, which results from measuring the star's position against that background on two occasions six months apart. The apparent displacement is caused by the Earth's orbital motion around the Sun. Parallax is in effect a measurement of distance by triangulation.

PARALLEL PROCESSING   Most computers operate with one processor carrying out arithmetic and other operations. In a parallel processing computer, as many as 256 processors operate simultaneously.

PARALLEL UNIVERSES   See MANY-WORLDS COSMOLOGY.

PARSEC   A measure of astronomical distance, equal to about three light-years.

PARTICLE PHYSICS   The physics of elementary particles.

PASSBAND   A specified range of wavelength in the electromagnetic spectrum. It provides a convenient window in which the brightness of an object may be measured.

PERFECT COSMOLOGICAL PRINCIPLE   See COSMOLOGICAL PRINCIPLE.

PHASE TRANSITION   A change of state such as occurs when, for example, a liquid boils or freezes. Phase transitions occurred in the early Universe as it cooled and its composition changed.

PHENOMENON   A sense impression or observation.

PHOTOMETRY   The measurement of the brightness of a star or other astronomical object. It is typically related to a specific range of wavelength, or passband.

PHOTON   A quantum (discrete quantity) of light. The energy of a photon determines its wavelength.

PLANET   During their formation, many stars are believed to form a circumstellar disk which, as in the case of the Sun, coalesces into a system of small orbiting bodies. Such objects are known as planets. Conventionally, a rocky body in its own orbit around the Sun but smaller than a few hundred kilometers across is called an asteroid.

PLANETARY NEBULA   A bright shell of fluorescent gas thrown off by a star during the last stage of its life. Such objects are highly distinctive and easily identifiable, and are used as distance indicators in the measurement of the Hubble constant.

PLATONISM   Philosophy derived from that of Plato. See REALISM, IDEALISM, RATIONALISM; compare EMPIRICISM.

POPPERIANISM   The philosophy of Karl Popper. Essentially it maintains that, contrary to the principle of verification favored by the logical positivists, propositions are scientific only if they are in principle falsifiable. No matter how many times a theory is verified, it could still be false and, indeed, unscientific. For instance, the ancient Egyptian idea that the star Sirius caused the Nile to flood is unscientific because it was unfalsifiable at that time. The fact that this idea was repeatedly verified in that, whenever the star appeared, the Nile flooded, did not make it scientific.

POSITIVISM   See LOGICAL POSITIVISM.

PRIMORDIAL BLACK HOLE   A black hole created in the very early Universe, either from the collapse of primordial density fluctuations or in the phase transition when the Universe condensed from one state to another. Since the conditions of their formation are very different, it is possible for primordial black holes to have far less mass than black holes formed from collapsing stars. Jupiter, for instance, could never become a black hole, but there could be primordial black holes of Jupiter's mass.

PRINCIPLE OF PARSIMONY   See OCCAM'S RAZOR.

PROPER MOTION   All stars have small random velocities relative to their neighbors. For nearby stars this can be seen as an displacement relative to the background of distant stars that becomes apparent over periods of, typically, a few years or more. This movement across the sky is known as proper motion.

PROTON   A positively charged particle with about the same mass as the neutron. Protons and neutrons are present in the nuclei of atoms in roughly equal numbers.

PROTOSTAR   A star in the process of being formed from a gas cloud.

PULSAR   A rapidly rotating neutron star with a strong magnetic field, which emits pulses of radio waves. A neutron star is a star in a very advanced state of its evolution that has collapsed to about 20 km in diameter. Further gravitational collapse is countered by internal pressure provided by the degenerate state of the star's material. Newly collapsed neutron stars rotate extremely rapidly and emit powerful beams of radio waves that sweep around like a lighthouse beam. It is these beams we observe as pulsars. Their pulse rates range from 4 seconds down to a few thousandths of a second— thus the fastest pulsars are known as *millisecond pulsars*.

PULSATING STAR   In the later stages of a star's life the outer atmosphere can become unstable and pulsate, causing the star to vary in brightness in a regular and characteristic way. Examples of such stars are RR Lyrae variables and Cepheid variables.

QUANTUM (plural QUANTA)   The indivisible unit in which energy may be emitted or absorbed.

QUANTUM FLUCTUATIONS   According to Heisenberg's uncertainty principle, variables such as energy and momentum can never be known exactly. Instead, they fluctuate within close, well-defined limits. According to the new theory of inflation, these quantum fluctuations could have produced small perturbations in the uniform density of the single space-time bubble that shaped our Universe.

QUANTUM MECHANICS (QUANTUM PHYSICS)   The theory, developed from Max Planck's quantum principle and Werner Heisenberg's uncertainty principle, dealing with the behavior of matter on the smallest scales. In this regime, the behavior of matter is often counterintuitive.

QUANTUM STATES   The state of a system as a network of potentialities with indefinite and hence unpredictable outcomes.

QUARK   One of a class of elementary particles that are the building blocks of protons and neutrons, each of which are composed of three quarks.

QUARK-HADRON TRANSITION   A change of state in the early Universe in which free quarks coalesced into atomic particles (hadrons) such as protons and neutrons.

QUARK NUGGETS   Hypothetical bodies formed from a catastrophic collapse of quarks in the early Universe. They have been postulated as a possible component of dark matter.

QUASAR (QUASI-STELLAR OBJECT)   The most luminous type of objects known in the Universe. Quasars are believed to be the compact luminous nuclei of galaxies, and to consist of a bright accretion disk surrounding a massive black hole. They have a starlike appearance, and their spectra are characterized by broad emission lines. Although most quasars are radio quiet, the first to be detected emitted radio waves and were consequently known, first as *radio stars,* and then as *quasi-stellar radio sources.*

RADIO ASTRONOMY   Many celestial objects emit radio waves. Radio astronomy is the study of the distribution and structure of these radio sources.

RADIOMETRIC DATING   Measurement of the age of objects like rocks by means of the decay of the unstable radioactive elements they contain. By this means it has been possible to calculate directly the ages of the Earth, the Moon, and meteorites, thus setting a lower limit to the age of the Universe.

RADIO SOURCE   Any astronomical object that emits radio waves and can be detected by radio telescopes.

RADIO TELESCOPE   A telescope designed to study cosmic radio sources. Many sophisticated radio telescope systems have been constructed, either using single receivers or several receivers combined in arrays.

RATIONALISM   A philosophy that stands in opposition to the doctrine of empiricism and is synonymous with realism and idealism. In rationalism, the only way to apprehend reality is through the power of the intellect.

RAY TRACING   A method of reconstructing optical images by tracing the paths of individual light rays in a computer.

REALISM   In this book, the term *realism* is used in its medieval sense of representing the position that regards Platonic forms, or "universals," as the only reality. It is therefore synonymous with idealism and rationalism, and opposed to empiricism. In Plato's philosophy, ultimate reality consists in ideal forms so that, for instance, a table is no more than an imperfect and ephemeral reflection of the real table that has its being in the realm of true reality, and whose essential nature can be apprehended only through the power of the intellect. See also IDEALISM, RATIONALISM.

RECESSION See EXPANSION OF THE UNIVERSE, VELOCITY OF RECESSION.

REDSHIFT The reddening of light from a star that is moving away from us, due to the DOPPLER EFFECT. It is measured by observing the shift of absorption or emission lines in the spectrum.

RELATIVITY, GENERAL THEORY OF A theory of gravity based on the principles of special relativity in which gravity is interpreted as the curvature of space and time.

RELATIVITY, SPECIAL THEORY OF The idea that the laws of science should be the same for all freely moving observers, no matter what their speed. This theory of space, time, and motion was formulated by Einstein in 1905, and in 1915 was generalized by him to include gravity.

RR LYRAE STAR A pulsating variable star with a characteristic luminosity. Their variations are utilized extensively in trying to determine the Hubble constant by means of the cosmic distance scale.

SCHMIDT PLATE A large photographic plate exposed in a Schmidt telescope, and covering an area on the sky of some 40 square degrees.

SCHMIDT TELESCOPE A type of telescope designed by Bernard Schmidt specifically to cover a very large field of view. Schmidt telescopes are used for surveys and large-scale statistical projects.

SECOND LAW OF THERMODYNAMICS See ENTROPY.

SEMANTIC Having to do with the references and hence the meaning of language, as opposed to its logical structure or syntax or grammar.

SEYFERT GALAXY A type of galaxy with a bright compact nucleus. Seyfert galaxies are now believed to be essentially low-luminosity quasars.

SINGULARITY In classical physics, a singularity is a state of infinite density where all matter and energy are compressed into a single infinitely small point. The curvature of space-time at this point is infinite, with the result that the concept of space-time ceases to have any meaning. In other words, it is a "place" where general relativity and all the other laws of physics break down. The big-bang singularity is the boundary condition of the Universe—the edge of space-time beyond which the existence or nonexistence of objects is a matter for metaphysics rather than physics. Similarly, because all scientific theories are formulated in terms of space and time, singularities within black holes represent a state of being, or nonbeing, which is beyond our comprehension, let alone scientific speculation.

SOLAR ECLIPSE See ECLIPSE.

SOLAR SYSTEM   The system of planets and their satellites orbiting the Sun, together with comets, asteroids, dust, and other debris.

SPACE TELESCOPE   An orbiting telescope remotely controlled from the Earth. Such telescopes are designed to observe radiation, such as X-rays, which cannot penetrate the Earth's atmosphere; or, like the Hubble Space Telescope, to observe at optical wavelengths free of the atmosphere's distorting effect.

SPACE-TIME   It is necessary to treat space and time on a similar footing in both the general and special theories of relativity. The ensuing mathematical space is called space-time.

SPECTROSCOPY   The study of the emission of electromagnetic radiation as a function of wavelength. In practice this means the analysis of the absorption and emission lines of atoms and molecules seen in the spectra of astronomical bodies.

SPECTRUM   The distribution of radiation as a function of wavelength. See SPECTROSCOPY.

SPIRAL GALAXY   A galaxy of, typically, between 10^{10} and 10^{11} times the mass of the Sun having the appearance of a spiral. Such galaxies comprise a thin disk with a central bulge surrounded by a diffuse halo of old stars. The disk has a spiral structure consisting of two to five arms winding out from the central bulge.

STAR   A concentration of gas large enough to shine by sustained nuclear fusion reactions. Stars are the most common visible components of galaxies. They form from fragments of collapsing gas clouds, which heat up sufficiently for thermonuclear reactions to commence. Young stars soon enter a long stable phase in which they burn hydrogen. Toward the end of their lives they undergo a complicated series of changes, depending on their mass. In general, the more massive the star, the hotter and more luminous it will become and the shorter its life. Below 8 percent of the Sun's mass, the condensing gas fragments never achieve a high enough temperature to burn hydrogen; such objects are known as brown dwarfs.

STAR CLUSTER   Stars are born from giant molecular clouds in small groups or larger clusters. Many of these clusters survive for considerable periods of time, and are useful to astronomers as distance indicators since their stars will all be at nearly the same distance from us, and usually born at the same time with the same composition. Clusters are broadly divided into two categories: the younger *open clusters* found in the galactic disk, and the much larger and more compact GLOBULAR CLUSTERS, which reside in the halo and are remnants of the formation of the Galaxy.

STATIC UNIVERSE   A universe in which the attractive force of gravity is exactly balanced by the repulsive force of a cosmological constant. Such a universe is now known to be too unstable to be sustainable.

STEADY-STATE COSMOLOGY   A cosmology proposed by Fred Hoyle, Hermann Bondi, and Thomas Gold in 1948. In a steady-state Universe, the perfect cosmological principle applies: on the large scale, a steady-state universe will appear the same to an observer in any position at any time. In other words, the Universe had no beginning and will have no end. The observed recession of distant galaxies is explained by the continuous creation of matter: as galaxies move farther away, new galaxies are formed in between them so that the Universe's average density is kept constant. This model has recently been revised so that creation takes place in intermittent bursts, causing a general expansion, which is followed by a gravitational collapse, which in turn gives rise to the next wave of creation and expansion. The steady-state model has the great scientific merit of making definite and testable predictions.

STELLAR EVOLUTION   The course of a star's life, from the moment of its birth in a gas cloud to its death when no further sources of energy are available to it.

STRANGE QUARK   Of the six quarks currently thought to be the most fundamental building blocks of matter, the strange quark is the third lightest, after the down and up quarks.

SUBATOMIC PARTICLES   Particles that are the building blocks of atoms, such as electrons and quarks.

SUPERGIANT STAR   In the later stages of stellar evolution, the outer part of a star's envelope expands by a very large amount, producing a huge increase in the star's luminosity. The most luminous of these stars are known as supergiants.

SUPERNOVA   The explosive death of a star. One type of supernova explosion occurs in a star that has used up all its nuclear fuel. It thus has no means of generating radiation in its core to counteract the weight of its outer layers, which suddenly collapse to form a neutron star or black hole. It is possibly the collision between these collapsing layers and the core which causes the supernova explosion. In another type of supernova, a white dwarf in a binary system can acquire sufficient mass from its partner to take it over the mass threshold for stability as a white dwarf, causing it to collapse to form a neutron star and resulting in a supernova explosion. Because of their great luminosity, 10^8 or even 10^{10} times as luminous as the Sun at maximum brightness, supernovae are used extensively as distance indicators for the measurement of the Hubble constant.

SUPERPHENOMENON   A phenomenon that is beyond sensory experience or observation, belonging to the realms of metaphysics.

TAUTOLOGY   Anything that is so trivially true that it is in effect meaningless. Examples of tautologies are; "The red ball is red," "2 = 2," and, arguably, all mathematical equations and other deductive statements since their conclusions are contained in their premises.

TELESCOPE   In the broadest sense, any instrument for gathering radiation. In most cases the radiation is focused to form an image that is then recorded by an appropriate detector. The most familiar are optical telescopes employing mirrors or lenses to focus light onto electronic detectors, but other telescopes exist or have been conceived, sometimes with radically different designs, to detect widely differing forms of radiation from neutrinos to gravitational waves.

TESTABILITY   See FALSIFIABILITY.

THEORY   In physics, a mathematical description of physical phenomena which can be compared with observations to establish its validity.

THOUGHT EXPERIMENT   An experiment that can be reasoned through by thought and intuition alone.

TIME   See ARROW OF TIME.

TIME DILATION   In general relativity, as a body recedes from an observer its time will appear to run more slowly, making events of a certain duration seem to take longer.

TOP QUARK   The most massive of the quark family. Such quarks may have seeded primordial black holes during the quark-hadron transition.

TRIANGULATION   The measurement of distance by geometrical methods.

ULTRAVIOLET EXCESS   Many astronomical objects, especially those undergoing cataclysmic processes, show an abnormal amount of ultraviolet radiation in their spectra. This is known as ultraviolet excess.

ULTRAVIOLET LIGHT   Light in the waveband adjacent to the optical region, with shorter wavelengths (about 10 to 300 mm). Since the atmosphere is opaque to these wavebands, the detection of ultraviolet radiation has to be conducted above the Earth's atmosphere.

UNCERTAINTY PRINCIPLE   The principle that one can never accurately measure both the position and the velocity of a particle since the more accurately the one is measured, the less accurate is the measurement for the other.

UNIVERSE   In principle, the Universe encompasses all that exists. However, theoretical cosmologies can imply realms of space and time with which we can have no conceivable contact. Their inclusion within our Universe is problematic. In this book, such universes are given a lowercase "u," as are mathematical "model" universes.

VARIABLE STAR   A star whose brightness varies over time. See ECLIPSE, PULSATING VARIABLE.

VELOCITY OF RECESSION   In the expanding Universe, galaxies are in general receding from one another. Their velocity of recession is proportional to their distance, the constant of proportionality being the HUBBLE CONSTANT.

VERIFICATION   The confirmation that the prediction of a theory is fulfilled.

VIENNA SCHOOL   A group of logical positivists strongly influenced by Ludwig Wittgenstein.

VIRGO CLUSTER   The nearest large cluster of galaxies to our Galaxy. The measurement of the distance to the Virgo Cluster is considered to be a fundamental step in the construction of the cosmic distance ladder and the measurement of the Hubble constant.

WAVELENGTH   The distance between two adjacent crests of waves. With light waves, the wavelength determines the color; thus red light has a longer wavelength than blue light.

WEAKLY INTERACTING MASSIVE PARTICLE (WIMP)   A generic term for slow, massive elementary particles that have been hypothesized as making up the missing matter. Such particles are nonbaryonic and would only rarely interact with ordinary atomic matter. They would thus be exceptionally difficult to detect.

WHITE DWARF   The final stage in the life of a star which is less than 1.4 times the mass of the Sun. The star collapses to a white-hot degenerate body that gradually cools to a dense dark cinder with a radius of about 1,000 km.

WIDE-FIELD ASTRONOMY   The technique of observing many astronomical objects simultaneously. Applications range from the use of wide-field photographs to fiber-fed spectrographs able to record many spectra simultaneously. Typically, wide-field astronomy tackles large-scale statistical problems.

WIMP   See WEAKLY INTERACTING MASSIVE PARTICLE.

X-RAY ASTRONOMY   Many classes of objects emit X-rays: stars, supernovae, and quasars, for example. X-ray astronomy, which detects such emissions, has to be carried out above the Earth's atmosphere, which is opaque to the X-ray wavebands.

BIBLIOGRAPHY

Alcock C. et al. "Possible gravitational microlensing of a star in the Large Magellanic Cloud." *Nature,* 365 (1993): 621.

Arp, Halton C., C. Roy Keys, and Konrad Rudnicki (eds.). *Progress In New Cosmologies: Beyond the Big Bang.* New York: Plenum Press, 1993.

Ayer, A. J. *The Central Questions of Philosophy.* London: Penguin, 1978.

Barrow, John D. *Theories of Everything.* Oxford: Clarendon Press, 1991.

Canizares, C. R. "Manifestations of a cosmological density of compact objects in quasar light." *Astrophysical Journal,* 263 (1982): 508.

Crawford, M., and D. N. Schramm. "Spontaneous generation of density perturbations in the early Universe." *Nature,* 298 (1992): 538.

Darwin, Francis (ed.). *Charles Darwin's Autobiography.* New York: Schuman, 1950.

Davies, Paul (ed.). *The New Physics.* Cambridge: Cambridge University Press, 1994.

Dawkins, Richard. *The Blind Watchmaker.* London: Penguin, 1988.

De Vaucouleurs, Gerard. *The Cosmic Distance Scale and the Hubble Constant.* Canberra, Mount Stromlo and Siding Springs Observatories, Australian National University, 1982.

Dicke, R. H., P. J. E. Peebles, P. G. Roll, and D. T. Wilkinson. "Cosmic blackbody radiation." *Astrophysical Journal,* 142 (1965): 414.

Edwards, Rob. "Rocket for star-struck astronomers." *Scotland on Sunday,* November 21, 1993, p. 4.

Ferris, Timothy. *Coming of Age in the Milky Way.* New York: William Morrow, 1988.

Freedman, W. et al. "Distance to the Virgo Cluster galaxy M100 from Hubble Space Telescope observations of Cepheids." *Nature,* 371 (1994): 757.

Gondhalekar, P. M. "Ultraviolet spectra of a large sample of quasars." *Monthly Notices of the Royal Astronomical Society,* 243 (1990): 443.

Guth, A. H. "Inflationary Universe: A possible solution to the horizon and flatness problems." *Physical Reviews D,* 23 (1981): 347.

Hamlyn, D. W. (trans.). *Aristotle's De Anima, Books II & III.* Oxford: Clarendon Press, 1978.

Hawking, Stephen W. *A Brief History of Time.* London: Bantam Press, 1988.

Hawkins, M. R. S. "Variable extragalactic objects: Identification and analysis of a complete sample to $B = 21$." *Monthly Notices of the Royal Astronomical Society,* 202 (1983): 571.

"Three unusual cataclysmic variables." *Nature,* 301 (1983): 688.

"Direct evidence for a massive galactic halo." *Nature,* 303 (1983): 406.

"A study of the galactic halo from a complete sample of RR Lyrae variables to $B = 21$." *Monthly Notices of the Royal Astronomical Society,* 206 (1984): 433.

"Gravitational microlensing, quasar variability and missing matter." *Nature,* 366 (1993): 242.

"Dark matter from quasar microlensing." *Monthly Notices of the Royal Astronomical Society,* 278 (1996): 787.

Hawkins, M. R. S., and Philippe Véron. "The evolution of the quasar luminosity function." *Monthly Notices of the Royal Astronomical Society,* 275 (1995): 1102.

"The space density of quasars at $z > 4$." *Monthly Notices of the Royal Astronomical Society,* 281 (1996): 348.

Horgan, John. "Karl R. Popper, the intellectual warrior." *Scientific American,* Nov. 1992, p. 20.

Hoyle, Fred. *The Black Cloud.* London: Heinemann, 1957.

The Intelligent Universe: A New View of Creation and Evolution. London: Michael Joseph, 1983.

Home Is Where the Wind Blows. Mill Valley, Calif.: University Science Books, 1994.

Hoyle, F., G. Burbidge, and J. V. Narlikar. "A quasi-steady-state cosmological model with creation of matter." *Astrophysical Journal,* 410 (1993): 437.

Hoyle, F., and W. A. Fowler. "Nature of strong radio sources." *Nature,* 197 (1963): 533.

Hume, David. *On Human Nature and the Understanding.* New York: Macmillan, 1978.

Iwasaki, Y., K. Kanaya, S. Kaya, S. Sakai, and T. Yoshie. "Nature of the finite temperature transition in QCD with strange quark." *Nuclear Physics B,* S42 (1995): 499–501.

Latour, Bruno, and Steve Woolgar. *Laboratory Life: The Social Contribution of Scientific Facts.* Beverly Hills and London: Sage Publications, 1979.

Layzer, David. *Cosmogenesis: The Growth of Order in the Universe.* New York: Oxford University Press, 1991.

Malpass, Brian. *Bluff Your Way in Science.* London: Ravette Books, 1993.

Matthews, Gwynneth. *Plato's Epistemology.* London: Faber & Faber, 1972.

Matthews, Robert. "Space hype." *Sunday Telegraph,* November 21, 1993, p. 19.

Montefiore, H. *The Probability of God.* London: SCM Press, 1985.

Paczynski, B. "Gravitational microlensing at large optical depth." *Astrophysical Journal,* 304 (1986): 1.

Penzias, A. A., and R. W. Wilson. "A measurement of excess antenna temperature at 4080 Mc/s." *Astrophysical Journal,* 142 (1965): 419.

Pickering, A. *Constructing Quarks: A Sociological History of Particle Physics.* Chicago: University of Chicago Press, 1984.

Plato. *The Last Days of Socrates.* London: Penguin, 1978.

Popper, Karl R. *Conjectures and Refutations.* London: Routledge & Kegan Paul, 1981.

  Objective Knowledge. Oxford: Oxford University Press, 1981.

Press, W. H., and J. E. Gunn. "Method for detecting a cosmological density of condensed objects." *Astrophysical Journal,* 185 (1973): 397.

Price, Lucien. *Dialogues of Alfred North Whitehead.* New York: Mentor, 1956.

Radford, Tim. "Cosmic mystery emerges from black hole." *Guardian,* November 18, 1993, p. 26.

Rees, M. J. "Black hole models for active galactic nuclei." *Annual Review of Astronomy and Astrophysics*, 22 (1984): 471.

Riordan, Michael, and David Schramm. *The Shadows of Creation: Dark Matter and the Structure of the Universe.* Oxford: Oxford University Press, 1993.

Sahu, K. C. "Stars within the large Magellanic Cloud as potential lenses for observed microlensing events." *Nature*, 370 (1994): 275.

Schild, R. E. "Microlensing variability of the gravitationally lensed quasar Q0957+561 AB." *Astrophysical Journal*, 464 (1996): 125.

Schild, R. E., and R. C. Smith. "Microlensing in the Q0957+561 gravitational mirage." *Astrophysical Journal*, 101 (1991): 813.

Schmidt, M. "3C 273: A star-like object with large red-shift." *Nature*, 197 (1963): 1040.

Schneider, P. "Upper bounds on the cosmological density of compact objects with sub-solar masses from the variability of QSOs." *Astronomy and Astrophysics*, 279 (1993): 1.

Schneider, P., J. Ehlers, and E. E. Falco. *Gravitational Lenses.* Berlin: Springer-Verlag, 1992.

Schneider, P., and A. Weiss. "A gravitational lens origin for AGN variability? Consequences for microlensing." *Astronomy and Astrophysics*, 171 (1987): 49.

Scott, P. F. et al. "Measurement of structure in the cosmic background radiation with the Cambridge Anisotropy Telescope." *Astrophysical Journal*, 461 (1996): L1.

Smoot, George, and Keay Davidson. *Wrinkles in Time.* London: Little, Brown, 1993.

Taylor, A. N., and M. Rowan-Robinson. "The spectrum of cosmological density fluctuations and nature of dark matter." *Nature*, 359 (1992): 396.

van den Brul, Caroline. "Perceptions of science: How scientists and others view the media reporting of science." Sixth Guardian Lecture, Nuffield College, Oxford, June 22, 1994.

Weinberg, Steven. *Dreams of a Final Theory.* London: Vintage Books, 1993.

 The First Three Minutes. London: Flamingo, 1993.

Wittgenstein, Ludwig. *Philosophical Investigations.* Oxford: Basil Blackwell, 1978.

 Tractatus Logico-Philosophicus. London: Routledge & Kegan Paul, 1978.

INDEX

Page numbers in **bold** refer to the main entry in the glossary

Wilson, Robert, 36, 186, 221

WIMPs. *See* weakly interacting
massive particles

Wittgenstein, Ludwig, 40–41, 43,
47–49, 85–86, 91, 186–187,
189, 205, 217, 222

Woolgar, Steve, 9, 185, 221

Woolley, Sir Richard, 19

X

X-ray astronomy, **217**

X-ray space telescopes, 120

X-rays, 80, 90, 118, 193, 200, 214,
217

Property of the
YORK COUNTY   LLEGE
112 C
Wells, M
(207) 646-9262